The Digital Mind

The Digital Mind

How Science Is Redefining Humanity

Arlindo Oliveira

The MIT Press
Cambridge, Massachusetts
London, England

Set in Stone Sans and Stone Serif by Toppan Best-set Premedia Limited. Printed and bound in the United States of America.

Cataloging-in-Publication information is available from the Library of Congress.

ISBN: 978-0-262-03603-0

10 9 8 7 6 5 4 3 2 1

to Ana Teresa

Contents

Foreword

Computers are as inert in their physical form as they are useless as thinking machines. Or so we might have thought a long time ago. Never the author of this book, however. For many years, decades now, he has been known to derail perfectly balanced conversations to point out signs that the same natural processes that drive the evolution of living systems are taking residence in the new digital realms. As time went by, the signs became more momentous and the outcome more strangely wonderful. Those who were startled by one of these derailed conversations (as I was) will, ever after, yearn for more. After all, Arlindo should know. His distinguished academic career straddles the whole gamut, from circuit design to machine learning, and increasingly so in a biological context. It is therefore not a surprise that he has a lot to say about the future of thinking machines. But, as the reader of this book will discover, there is more here—a lot more.

As a student in computer science at Berkeley, Arlindo also completed a minor in neurosciences. At some point I congratulated him for the foresight of what looked to me like a solid interdisciplinary foundation for his subsequent achievements. He characteristically dismissed this minor as a "not very useful" impulse of his youth. Indeed, even then, he already had questions about digital minds that would just not go away. The answers about cognition he got then simply did not compute, quite literally. As you will find throughout this volume, the author is happy to speculate, but always within clear sight of computable explanations and historic context. And I don't mean just computable in principle, there has to be a trail back to the component circuitry. As a biologist, this compulsion is familiar territory. The understanding of complex processes invariably includes components at multiple scales, from genes, to cells and all the way to ecosystems. The same systemic discipline guides this voyage toward digital minds. Artful storytelling alone does not cut it. It would be "not very useful."

So we now find ourselves upheaved by a digital storm that pervades everything we sense and do. The new gadgetry, and the cloud that supports it, confine how we communicate as much as they expand how broadly we reach. More importantly, in our more mindful moments, this digital machinery enables us to see previously unfathomed horizons. The neurosciences now have a lot more to say about how our minds compute, by anticipating sensorial input with a mixture of innate and experientially learned models. And as one would expect, thinking about sensing the sensor strains our minds just as recursion challenges our digital machines. The boundaries between organic and inorganic are clearly no longer as divisive when it comes to mindful computation. These are the moments when one looks for these conversations with Arlindo, once more, maybe with the pretext of a game of Chess or Go (don't hold your hopes too high!). These are the questions he has been probing and practicing for decades in the computational realm. He should write a book about it. …

The last paragraph could have been a good ending for my foreword. However, this book has two surprising final chapters about what we are getting ourselves into. They go one step beyond what digital consciousness may be and how inevitably the process will unfold. This book is organized chronologically, as a novel, so I should not spoil the ending here. As in any good book, along the way we collect answers to our own questions, not just the author's. I, for one, couldn't resist seeing the emergence of digital creatures, and of digital minds, as part of the same homeostatic drive that propels biological evolution into ever more complex constructs. What better way to build digital machines that can handle the vagaries of a complex world than to have them also able to think it over. Digital or not, minds may just be the inevitable outcome, as the book explains. The physicist may instead be tempted to see symptoms of the informational nature of more atomic vehicles. Either way, our digital present looks ever more as the seed to a more mindful digital future.

One last word, this time of caution. If you have teenagers at home, make sure to finish the book before revealing its existence to the rest of the family. In more ways than we may be comfortable with, this book was really written for a new generation. If my own children offer some guidance, they may find in this volume the missing explanation for how we got them into this ever digitizing predicament. In that case, they will be as perplexed as we are about how oblivious we have been to the strangely wonderful and inevitable outcome. In other words, you may have trouble getting the book back. Maybe tell them there is a blog too.

Jonas Almeida

Preface

The seemingly disparate subjects of computer science, biology, and neuroscience have always fascinated me and, in one way or another, have always influenced my life and career choices. The attentive reader will notice that this book is, mainly, a collage of many ideas about these subjects—ideas I have encountered during the past three decades. The majority of these ideas come from the many talented writers, scientists, and philosophers I have met, sometimes in person but mostly through their works.

The book is somewhat redundant in a way; to understand the topics covered, you could go directly to the original sources. Nonetheless, I believe not many people have walked through life exactly the way I have, and have not had the opportunity I have had to encounter some of these fascinating places, thinkers, and concepts.

I became attracted to math and science early in my life. I always wanted to know how things worked, and studying science looked like an easy way to make a living doing exactly what I wanted to do. As a young boy, I was fascinated by science. I spent years playing with chemistry sets. The concept of a "safe chemistry set" had not yet been invented, and my days were spent concocting new ways to combine chemicals in order to obtain explosive or otherwise dangerous combinations. Flasks, test tubes, beakers, and other chemical apparatus would break, collapse, explode, or simply be made useless by our unguided and exploratory approach to chemistry, but the process provided me and my friends with an endless source of recreation. Furthermore, it looked like science to us.

A few years later, after buying some books and magazines devoted to do-it-yourself electronics, I assembled my first electronic circuits by the blind and laborious process of painstakingly buying individual electronic components and assembling them in accordance with nearly incomprehensible diagrams drawn by people who, unlike me, knew what they were doing.

In the late 1970s, it became possible to acquire a very primitive version of a personal computer. After acquiring a few even more primitive machines, I became the proud owner of a Sinclair ZX Spectrum. The ZX Spectrum, a small personal computer with 48 kilobytes of memory and an integrated keyboard, was a prodigy of the technology at the time. It used a television set as a display. It could be programmed in a dialect of the Basic programming language. It had no permanent memory, and any new program had to be saved to tape on a standard tape recorder. As often as not, depending on the quality of the tape and that of the recorder, I would fail to recover the saved program, but that didn't deter me from spending my days programming it. Working with the ZX Spectrum was my introduction to programming, and I was hooked on computers for life.

After studying electrical engineering at Técnico in Lisbon, I specialized in digital circuit design, a field that was in its infancy at the time and that developed in close parallel with many other communication and information technologies. In particular, I worked on very-large-scale integration (VLSI) circuits, a general term that is applied to integrated circuits with many devices. Under the auspices of a very influential scientist and politician, José Mariano Gago, Portugal had just signed an agreement with the European Organization for Nuclear Research (CERN). That agreement provided me with an opportunity to spend a summer internship in Geneva. There, at CERN, I tested my programming skills on the equipment that controlled the high-energy particle beams used for experiments in particle physics and was able to share in the unique enthusiasm of the physics community that was searching for the laws that govern the universe. My passion for particle physics was only made stronger by that summer in Geneva.

After my stint at CERN, I felt a need to be closer to the action in the rapidly moving field of digital circuits. I applied to, and was accepted at, the University of California at Berkeley, an institution that excelled in the development of new tools, circuits, and technologies for designing integrated circuits. There I made contact with a whole new world of techniques for designing integrated circuits, many of them developed in the Computer Aided Design (CAD) group then led by Alberto Sangiovanni-Vincentelli (my advisor), Robert Brayton, and Richard Newton. But Berkeley, a school with so many top people in so many fields, made it possible for me to become familiar with a number of other areas that had been relatively unknown to me. Two of these areas fascinated me and defined my future interests: algorithms and neuroscience.

I learned algorithms at Berkeley with Richard Karp, one of the founding fathers of algorithms and complexity theory. Creating algorithms that could be used in designing integrated circuits was the mission of the Berkeley CAD group. Designers of integrated circuits use algorithms to create, verify, simulate, place, and connect the transistors, the working units of all electronic gadgets. Without those complex algorithms, none of today's integrated circuits could have been designed and fabricated.

At Berkeley, I also had the chance to learn about another field that soon would have many connections with algorithms and computers: neuroscience. I took a minor in neuroscience and managed to learn just enough about evolution, neurons, and brains to squeeze by, but the little I learned was enough to make neuroscience another passion of mine. A number of computer scientists had begun to develop algorithms with which to process biological data, a field that would become later known as bioinformatics. Because of my interests, developing algorithms to address biological problems was an obvious choice for me. Over the years, I was able to make some contributions in that field.

Recently I had an opportunity to review proposals in the scope of the Human Brain Project. The proposals covered many fields, from brain simulation to bioinformatics and brain therapy, and made me more aware of the importance of the connections between the fields. The Human Brain Project aims at using computers to support brain research. In a way, I felt that, with the application of computers to the central problem of understanding the human brain, my trip through science had come full circle.

That trip was influenced by many people. My advisors, colleagues, and students accompanied me, worked with me, and were instrumental in the few relevant scientific results I was able to obtain. The colleagues who worked with me are now spread around the world. My exchanges with them have influenced me in ways that are impossible to describe fully.

However, I believe that the books I read during these decades probably were my greatest influences. Several of them changed the way I saw the world so profoundly that my life would certainly have been very different had I not encountered them. Certainly the book you are reading now would not have been possible were it not for the many authors who, before me, explored the relations between computation, evolution, intelligence, and consciousness.

Although I was influenced by many, a few authors deserve special and explicit mention here. Richard Dawkins and Stephen Jay Gould opened my mind to the wonders of evolution. Daniel Dennett and Douglas Hofstadter steered me toward working in artificial intelligence and developed my interest in the problems of brain, mind, and consciousness. David Hubel described so clearly the way some areas of the brain are organized that one gets the feeling that with a little more effort he could have also explained the behavior of the whole mind. Steven Pinker, Marvin Minsky, and Roger Penrose influenced deeply my own views about the workings of the human mind and the meaning of intelligence, even though I disagree with them in some respects. Kevin Kelly's ideas on the future of technology and Ray Kurzweil's unwavering belief in the technological singularity have also been strong influences. James Watson and John Craig Venter provided me with unique insights into how the biological sciences have progressed in the past fifty years. Eric Drexler's vision of the future of nanotechnology changed my view of the very small and of the world. Sebastian Seung's, Olaf Sporns', and Steven Rose's descriptions of their involvement in projects that aim at understanding the brain were enlightening and inspirational. Nick Bostrom addresses many of the matters covered herein, but develops them further in ways I did not even dream of. Jared Diamond and Yuval Harari changed deeply the way I view human history and human evolution.

Other strong influences come from science fiction, of which I am an avid reader (too avid, some readers will say). Isaac Asimov, Arthur C. Clarke, and Robert Heinlein created in me a lasting interest for science fiction. Greg Egan, Vernor Vinge, Neal Stephenson, Larry Niven, and Charles Stross propose such clear and challenging visions of the future that it becomes easy to believe they will one day happen, strange and perturbing as they are.

I have tried to make this book easy to follow for anyone interested in the topics it addresses. The book is aimed at readers with a general interest in science and technology, and no previous knowledge of any of the many areas covered should be necessary. The final three chapters include no technical material. They present the book's central argument and can be read independently.

Readers with less technical backgrounds should not be discouraged by the occasional equations, diagrams, and mathematical arguments. I tried to include enough information to help readers grasp some of the technical details, but in most cases the central ideas can be gathered from the accompanying explanations and no significant information will be lost by skipping the details.

Acknowledgments

I thank everyone who supported me and helped me write this book, from the many authors who inspired it to my family, including my daughters, Ana Sofia and Helena, Mariana and Clara, and my parents, Arlindo and Amália. I hope they will, one day, be able to read this book and to understand some of the more radical ideas set forth in it. I fear, however, that for my parents this may turn out not to be the case. Given the nature of science and technology, we become rapidly obsolete and unable to understand new developments. Sometimes things change just too fast for us to follow them.

I also thank the institutions at which I was privileged to work, including Instituto Superior Técnico, the University of California at Berkeley, Cadence Design Systems, CERN, and INESC. They were, for decades, my home away from home.

Someone has said a book is never finished, only abandoned by its author. I abandoned this book at the end of my recovery from a fairly serious surgical intervention. The time I needed for the recovery enabled me to bring to fruition what had been, until then, little more than a dream. I owe this possibility to the many friends who helped me and also to the talented surgeon who operated on me, Dr. Domingos Coiteiro.

The overall organization of the book and the actual text owe much to the many friends and colleagues who patiently read and commented on drafts. I would like to thank, in particular, Jonas Almeida, Miguel Botto, Luís Pedro Coelho, Tiago Domingos, Mário Figueiredo, Alexandre Francisco, Vítor Leitão, Pedro Guedes de Oliveira, Amílcar Sernadas, Manuel Medeiros Silva, Larry Wald, and the anonymous reviewers for many useful suggestions and corrections. I also thank Miguel Bugalho, Afonso Dias, Patrícia Figueiredo, Alexandre Francisco, Bobby Kasthuri, and Richard Wheeler for the material they contributed for use in the book.

I thank Pedro Domingos for suggesting the MIT Press as the right publisher, Marie Lufkin Lee for believing in the project, and Kathleen Hensley for handling all the technical matters related to the preparation of the manuscript. I am especially indebted to Paul Bethge, the manuscript editor, who greatly contributed to the readability of this book with his suggestions, corrections, and improvements.

During the writing, as well as during the years that preceded it, my wife, Ana Teresa, was my support and my strength, with a dedication that was clearly above and beyond the call of duty. I thank her from the bottom of my heart, and dedicate this book to our future together.

1 The Red Queen's Race

To a casual observer, computers, cells, and brains may not seem to have much in common. Computers are electronic devices, designed by humans to simplify and improve their lives; cells, the basic elements of all living beings, are biological entities crafted by evolution; brains are the containers and creators of our minds, with their loads of hopes, fears, and desires.

However, computers, cells, and brains are, in one way or another, simply information-processing devices. Computers represent the latest way to process information, in digital form. Before them, information processing was done by living organisms, in effect creating order out of chaos. Computers, cells, and brains are the results of complex physical and chemical processes as old as the universe, the most recent products of evolution, the winners of a race started eons ago.

Everything Is Connected

We are made of the atoms created in the Big Bang, 14 billion years ago, or in the explosions of distant stars that occurred for billions of years after that event, which signaled the beginning of time. At every breath, every one of us inhales some of the very same oxygen molecules that were in the last breath of Julius Caesar, reused over the centuries to sustain all life on Earth. We know now that the wing beat of a butterfly may indeed influence the path of a hurricane many months later, and that the preservation of one species may be critically dependent on the life of all other species on Earth. The evolutionary process that started about 4 billion years ago has led to us, and to almost everything surrounding us, including the many devices and tools we own and use.

This book touches many different areas I view as sharing strong connections. It covers computers, evolution, life, brains, minds, and even a bit of physics. You may view these areas as disjoint and unrelated. I will try to show you that they are connected and that there is a common thread connecting physics, computation, and life. I am aware that the topic of each of the chapters of this book fully deserves a complete treatise by itself (in some cases, several treatises). I hope the coherence of the book is not affected by the fact that each of the many areas covered is touched upon only lightly, barely enough to enable the reader to understand the basic principles.

In a way, everything boils down to physics. Ernest Rutherford supposedly said that "science is either physics or stamp collecting," meaning that the laws of physics should be sufficient to explain all phenomena and that all other sciences are no more than different abstractions of the laws of physics. But physics cannot be used to directly explain or study everything. Computation, biology, chemistry, and other disciplines are necessary to understand how the universe works. They are not unrelated, though. The principles that apply to one of them also apply to the others.

This realization that everything is related to everything else is, in large part, a result of our improved understanding of the world and our ability to master technology. Only relatively recently have advances in science made it possible for us to understand that the same equations describe the light we receive from the sun, the behavior of a magnet, and the workings of the brain. Until only a few hundred years ago, those were all independent realities—separate mysteries, hidden from humanity by ignorance.

Technology, by making use of the ever-improving understanding provided by science, has been changing our lives at an ever-increasing pace for thousands of years, but the word *technology* is relatively new. Johann Beckmann, a German scholar, deserves credit for coining the word (which means "science of craft") in 1772, and for somehow creating the concept (Kelly 2010). Before Beckmann, the various aspects of technology were individually known as *tools*, *arts*, and *crafts*. Beckmann used the word *technology* in a number of works, including a book that was later translated into English as *Guide to Technology, or to the knowledge of crafts, factories and manufactories*. However, the word was rarely if ever used in common language before the twentieth century. In the second half of the twentieth century, use of the word in common language increased steadily. Today it is a very common word in political, social, and economic texts.

With technology came the idea that innovation and novelty are intrinsic components of civilization. Constant changes in technologies, society, and economics are so ingrained in our daily lives that it is hard to understand that this state of affairs wasn't the rule in the ancient days. A few hundred years ago, change was so slow that most people expected the future to be much like the past. The concept that the future would bring improvements in people's lives was never common, much less popular. All that changed when changes began to occur so often that they were not only perceptible but expected. Since the advent of technology, people expect the future to bring new things that will improve their daily lives. However, many of us now fear that the changes may come too fast, and may be too profound, for normal people to assimilate them.

Old Dogs, New Tricks

If you are more than a few decades old, you probably feel that technology is changing so fast than you can't keep up with it, that new devices and fads of very dubious interest appear every day, and that it is hard to keep up with the pace of change. The things young people do and use today are increasingly foreign to the elders, and it is difficult to keep up with new trends, tools, and toys. Young children are better than you with a smartphone, are more at ease with computers, play games you don't understand, and flock to new sites that are of little interest to you. They don't even know what a VCR is—that miracle technology of the 1980s, which was famously difficult to operate. Even CDs and DVDs—digital technologies that emerged in the last few decades of the twentieth century—seem to be on their way out, having lasted for only a few decades (significantly less time than the vinyl records and film reels they replaced, which lasted almost a hundred years).

If you were born in the present century, on the other hand, you don't understand how your parents and grandparents can have such a hard time with technological innovations. New ideas and gadgets come naturally to you, and you feel at home with the latest app, website, or social network.

Yet reasonably literate and even technologically sophisticated people of my generation don't feel at all ready to give up on advancing technology. Our generation invented computers, cell phones, the World Wide Web, and DNA sequencing, among many other things. We should be able to understand and use any new things technology might throw at us in the next decades.

I believe that technology will keep changing at an ever-increasing pace. As happened with our parents and our grandparents before us, our knowledge about technology is likely to become rapidly obsolete, and it will be difficult to understand, use, and follow the technological developments of coming decades. There is some truth to the saying "you can't teach an old dog new tricks."

Whether our children and grandchildren will follow the same inevitable path to obsolescence will depend on how technology continues to change. Will new technological developments keep coming, faster and faster? Or are we now in a golden age of technological development, a time when things are changing as rapidly as they ever will?

When I was eight years old, I enjoyed visiting my grandfather. He lived in a small village located in what I then believed to be a fairly remote place, about sixty miles from Lisbon, the capital of Portugal. The drive from my house to the village where he lived would take us half a day, since the roads were narrow and winding. He would take me with him, to work in the fields, on a little wagon pulled by a horse. To a little boy, being able to travel on a horse-drawn wagon was quite exciting—a return to the old days. Other things would reinforce the feeling of traveling back to the past. There were no electric lights in his house, no refrigerator, no television, and no books. Those "modern" technologies weren't deemed necessary, as my grandparents lived in a centuries-old fashion.

By comparison, my parents, who left their village when they married, were familiar with very advanced technologies. They were literate, had a TV, and even owned a car. To my grandparents, those technologies never meant much. They had no interest in TV, newspapers, or books, most of which reported or referred to a reality so remote and so removed from their daily experience that it meant nothing to them. Cars, trains, and planes didn't mean a lot to people who rarely traveled outside their small village and never felt the desire to do so.

Forty years later, my parents still have a TV, still read books, and still drive around in a car. I would have thought that, from their generation to ours, the technological gap would have become much smaller. I would have imagined that, forty years later, they would be much closer to my generation than they were to the previous generation in their ability to understand technology. However, such is not the case, and the gap seems to increase with each generation. My parents, which are now over 80, never quite realized that a computer is a universal machine that can be used to play games, to obtain information, or to communicate, and they never understood that a computer is just a terminal of a complex network of information devices that can deliver targeted information when and where one needs it. Many people of their generation never understood that—except for minor inconveniences, caused by limitations in technologies, that will soon disappear—there is no reason why computers will not replace books, newspapers, radio, television, and almost every other device for the delivery of information.

You may think that my generation understands what a computer can do and is not going to be so easily outpaced by technological developments. After all, almost all of us know exactly what a computer is, and many of us even know how a computer works. That knowledge should give us some confidence that we will not be overtaken by new developments in technology, as our parents and our grandparents were. However, this confidence is probably misplaced, mostly because of the Red Queen effect, the name of which is inspired by the character in Lewis Carroll's masterpiece, *Through the Looking Glass*: "Now, here, you see, it takes all the running you can do, to keep in the same place."

The Red Queen effect results from the fact that, as evolution goes by, organisms must become more and more sophisticated, not to gain competitive advantage, but merely to stay alive as the other organisms in the system constantly evolve and become more competitive.

Although the Red Queen effect has to do with evolution and the competition between species, it can be equally well applied to any other environment in which competition results in rapid change—for example, business or technology.

I anticipate that, in the future, each generation will be more dramatically outpaced than the generation before it. Thirty years from now, we will understand even less about the technologies of the day than our parents understand about today's technologies. I believe this process of generational obsolescence will inevitably continue to accelerate, generation after generation, and that even the most basic concepts of daily life in the world of a hundred years from now would be alien to members of my generation.

Arthur C. Clarke's third law states that any sufficiently advanced technology is indistinguishable from magic. A hundred years from today, technology may be so alien to anyone alive today as to look like magic.

From Computers and Algorithms to Cells and Neurons

New technologies are not a new thing. However, never before have so many technological innovations appeared in such a short period of time as today. In coming decades, we will continue to observe the rapid development and the convergence of a number of technologies that, until recently, were viewed as separate.

The first of these technologies, which is recent but has already changed the world greatly, is computing technology, which was made possible by the development of electronics and computer science. Computers are now so pervasive that, to many of us, it is difficult to imagine a world without them. Computers, however, are useful only because they execute programs, which are nothing more than implementations of algorithms. Algorithms are everywhere, and they are the ultimate reason for the existence of computers. Without algorithms, computers would be useless.

Developers of algorithms look for the best way to tell computers how to perform specific computations efficiently and correctly. Algorithms are simply very detailed recipes—sequences of small steps a computer executes to obtain some specific result. One well-known example of an algorithm is the algorithm of addition we all learned in school. It is a sequence of small steps that enables anyone following the recipe to add any two numbers, no matter how large. This algorithm is at the core of each modern computer and is used in every application of computers.

Algorithms are described to computers using some specific programming language. The algorithms themselves don't change with the programming language; they are merely sequences of abstract instructions that describe how to reach a certain result. Algorithm design is, in my view, one of the most elegant and fascinating fields of mathematics and computer science.

Algorithms are always developed for specific purposes. There are many areas of application of algorithms, and two of these areas will play a fundamental role in the development of future technologies.

The first of these areas is machine learning. Machine learning algorithms enable computers to learn from experience. You may be convinced that computers don't learn and that they do only what they are explicitly told to do, but that isn't true. There are many ways in which computers can learn, and we use the ability of computers to learn when we watch TV, search the Web, use a credit card, or talk on a phone. In many cases the actual learning mechanisms are hidden from view, but learning takes place nonetheless.

The second of these areas is bioinformatics, the application of algorithms to the understanding of biological systems. Bioinformatics (also known as computational biology) uses algorithms to process the biological and medical data obtained by modern technologies. Our ability to sequence genomes, to gather data about biological mechanisms, and to use those data to understand the way biological systems work depends, in large part, on the use of algorithms developed specially for that purpose. Bioinformatics is the technology that makes it possible to model and understand the behavior of cells and organisms. Recent advances in biology are intricately linked with advances in bioinformatics.

Evolution, the process that has created all living things is, in a way, also an algorithm. It uses a very different platform to run, and it has been running for roughly 4 billion years, but it is, in its essence, an algorithm that optimizes the reproductive ability of living creatures. Four billion years of evolution have created not only cells and organisms, but also brains and intelligent beings.

Despite the enormous variety of living beings created by evolution and the range of new technologies that have been invented, members of the genus *Homo* have been the exclusive owners of higher intelligence on Earth. It is this particular characteristic than has enabled humans to rule the Earth and to adapt it to their needs and desires, sometimes at the expense of other important considerations.

Technology, however, has evolved so much that now, for the first time, we face the real possibility that other entities—entities created by us—could become intelligent. This possibility arises from the revolution in computing technologies that has occurred in the past fifty years, including artificial intelligence and machine learning, but also from the significant advances in our understanding of living beings—particular in our understanding of the human body and the human brain. Computing technologies, which are only a few decades old, have changed so many things in our daily lives that civilization as we now know it would not be possible without computers. Physics and biology have also made enormous advances, and for the first time we may have the tools and the knowledge to understand in detail how the human body and the human brain work.

Advances in medical techniques have already led to a significant increase in the life expectancy of most people alive today, but in coming decades we are likely to see unprecedented improvements in our ability to control or cure deadly diseases. These improvements will result from our increased understanding of biological processes—an understanding that will be made possible by new technologies in biology, physics, and computation. Ultimately, we may come to understand biological processes so well that we will be able to reproduce and simulate them in computers, opening up new possibilities in medicine and engineering.

In this book we will explore the possibility that, with the advances in medical and computing technologies, we may one day understand enough of the way the brain works to be able to reproduce intelligence in a digital support—that is, we may be able to write a program, executed by a digital computer, that will exhibit intelligence. There are a number of ways in which this could happen, but any one of them will lead to a situation in which non-biological minds will come into existence and become a members of our society. I called them *digital minds* because, almost certainly, they will be made possible by the existence of digital computer technology. The fact that non-biological minds may soon exist on Earth will unleash a social revolution unlike any that has been witnessed so far. However, most people are blind not only to the possibility of this revolution but also to the deep changes it will bring to our social and political systems. This book is also an attempt to raise the public awareness of the consequences of that revolution.

The majority of the predictions I will make here are likely to be wrong. So far, the future has always created things stranger, more innovative, and more challenging than what humans have been able to imagine. The coming years will be no exception and, if I am lucky enough to be alive, I will find myself surprised by new technologies and discoveries that were not at all the ones I expected. However, such is the nature of technology, and making predictions is always hard, especially about the future.

It is now time to embark on a journey—a journey that will take us from the very beginnings of technology to the future of the mind. I will begin by showing that exponential growth is a pattern built into the scheme of life but also a characteristic of the development of many technologies.

2 The Exponential Nature of Technology

Since the dawn of mankind, cultures and civilizations have developed many different technologies that have changed profoundly the way people live. At the time of its introduction, each technology changed the lives of individuals, tribes, and whole populations. Change didn't occur at a constant pace, though. The speed at which new technologies have been developed and used has been on the increase ever since the first innovations were introduced, hundreds of thousands of years ago.

Technological innovations are probably almost as old as the human species, and it is even possible some non-human ancestors used basic technologies more than 3 million years ago (Harmand et al. 2015). However, the rate at which new technologies have been introduced is far from constant. In prehistoric times, it took tens or even hundreds of thousands of years to introduce a new technology. In the last thousand years, new technologies have appeared at a much faster rate. The twentieth century saw new technologies appear every few years, and since then the pace has increased.

It is in the nature of technology that new developments are based on existing ones and, therefore, that every new significant development takes less time and effort than previous ones. This leads to an ever-increasing rate of technological change that many believe to be exponential over long periods as new and significant technologies are developed and replace the old ones as the engine of change (Bostrom 2014).

Prehistoric Technologies

Although stone and wood tools have been developed many times and are used by a number of species, we may consider that the first major technological innovation created by humans was the discovery (more precisely, the controlled use) of fire. Although the exact date for that momentous event is disputed, it probably happened between the lower and the middle portions of the Paleolithic period, between more than a million years and 400,000 years ago, and thus predated *Homo sapiens*.

The controlled use of fire must have brought a major change in the habits of prehistoric humans (Goudsblom 1992). Newer generations got used to the idea that meat would be easier to eat if grilled, and that some vegetables would taste better if cooked. Cooking makes food easier to digest and makes it possible to extract more calories from the available food supply (Wrangham 2009). The control of fire may have been the major reason for the increases in the sizes of humans' brains.

From our point of view, it may seem as if not much happened for thousands of years, until the domestication of plants and animals. That, of course, is not true. Significant cultural changes happened in the intervening years, including what Yuval Harari (2014) has called the cognitive revolution—a change that enabled humans to use language to communicate abstract thoughts and ideas. A large number of major technological breakthroughs happened in the thousands of years before the agricultural revolution. Bows and arrows, needles and thread, axes, and spears all were probably invented more than once and changed the societies of the time. The exact dates of these inventions or discoveries are mostly unknown, but they all happened probably between 50,000 and 10,000 years ago. One major invention every 10,000 years (roughly 400 generations) doesn't look like a breathtaking pace of change, but one must remember that hundreds of thousands of years elapsed between the time our ancestors descended from the trees and the time when any major changes in their habits occurred.

Roughly 10,000 to 11,000 years ago, humans had spread to most of the continents and had begun to live in villages. A few thousand years later, many cultures had begun to domesticate plants and animals. This represented what can be viewed as the second major technological revolution, after the controlled use of fire. Usually called the agricultural revolution, it deeply changed the way people lived and interacted. The change from the hunter-gatherer nomad way of life to sedentary agriculture-based living wasn't necessarily for the better. With the population increasing, humans became more dependent on a fixed supply of food, and larger communities led to an increase in the incidence of contagious diseases and to battles for supremacy. Still, these changes must have occurred, in most places, over a large number of generations, and, most likely, they have not caused a profound change in the way of life of any particular individual.

The speed of technological evolution has not stopped increasing since then. Although the exact date is disputed, the first wheel probably appeared in Mesopotamia around 3500 BC. Surprisingly enough, several cultures as recent as the Aztecs, the Mayans, and the Incas didn't use the wheel—probably owing to a lack of convenient draft animals. However, the majority of occidental and oriental cultures used that sophisticated implement extensively.

Many other inventions aimed at saving physical labor appeared during the past 10,000 years. Technological progress wasn't uniform in all civilizations or in all parts of the world. Although the exact reasons are disputed, there is a good case to be made that minor differences in the timing of the initial technological developments were exacerbated by further technological advances (Diamond 1997) and led to very different levels of development. Relatively minor environmental differences, such as the availability of draft animals, the relative ease of travel between places, and the types of readily available crops, led to differences in the timing of the introduction of technological developments, such as agriculture, transportation, and weapons. Those differences, in turn, resulted in differences in the timing of the introduction of more advanced developments—developments that were correlated with the more basic technologies, although sometimes in a non-obvious way. Information-based developments such as counting and writing, and other societal advances resulted from technological developments related to the more fundamental technologies. In the end, the cultures that had a head start became dominant, which eventually resulted in the dominance of Western society and culture, supported in large part by superior technological capacity. It isn't hard to see that civilizations which started later in the technological race were at a big disadvantage. Minor differences in the speed or in the timing of introduction of technological developments later resulted in huge differences. Western civilization ended up imposing itself over the entire known world, with minor exceptions, while other developed civilizations were destroyed, absorbed, or modified in order to conform to Western standards—a process that is still going on.

The First Two Industrial Revolutions

The industrial revolution of the late eighteenth century is the best-known discontinuity in technological development. Technological innovations in industry, agriculture, mining, and transportation led to major changes in the social organization of Britain, and, later on, of Europe, North America, and the rest of the world. That revolution marked a major turning point in human history, and changed almost all aspects of daily life, greatly speeding up economic growth. In the last few millennia, the growth of the world economy doubled every 900 years or so. After the technological revolution, the growth rate of the economy increased; economic output now doubles every 15 years (Hanson 2008).

That industrial revolution is sometimes separated into two revolutions, one (which took place in the late eighteenth century) marked by significant changes in the way the textile industry operated and the other (which began sometime in the middle of the nineteenth century) by the development of the steam engine and other transportation and communication technologies, many of them based on electricity. These revolutions led to a profound change in the way consumer products were produced and to widespread use of such communication and transportation technologies as the telephone, the train, the automobile, and the airplane.

The first industrial revolution began in Great Britain, then spread rapidly to the United States and to other countries in Western Europe and in North America. The profound changes in society that it brought are closely linked to a relatively small number of technological innovations in the areas of textiles, metallurgy, transportation, and energy.

Before the changes brought on by the first industrial revolution, spinning and weaving were done mostly for domestic consumption in the small workshops of master weavers. Home-based workers, who did both the spinning and the weaving, worked under contract to merchant sellers, who in many cases supplied the raw materials. Technological changes in both spinning and weaving brought large gains in productivity that became instrumental in the industrial revolution.

A number of machines that replaced workers in spinning, including the spinning jenny and the spinning mule, made the production of yarn much more efficient and cheap. Weaving, which before the revolution had been a manual craft, was automated by the development of mechanical looms, which incorporated many innovations, such as the flying shuttle. This led to a much more efficient textile industry and, with the invention of the Jacquard loom, to a simplified process of manufacturing textiles with complex patterns. The Jacquard loom is particularly relevant to the present discussion because it used punched cards to control the colors used to weave the fabric. Later in the nineteenth century, punched cards would be used in the first working mechanical computer, developed by Charles Babbage.

Whereas the first industrial revolution was centered on textiles and metallurgy, the second was characterized by extensive building of railroads, widespread use of machinery in manufacturing, increased use of steam power, the beginning of the use of oil, the discovery and utilization of electricity, and the development of telecommunications. The changes in transportation technology and the new techniques used to manufacture and distribute products and services resulted in the first wave of globalization, a phenomenon that continues to steamroll individual economies, cultures, and ecologies.

However, even the changes brought by improved transportation technologies pale in comparison with the changes brought by electricity and its myriad uses. The development of technologies based on electricity is at the origin of today's connected world, in which news and information travel at the speed of light, creating the first effectively global community.

During the first two industrial revolutions, many movements opposed the introduction of new technologies on the grounds that they would destroy jobs and displace workers. The best-known such movement was that of the Luddites, a group of English workers who, early in the nineteenth century, attacked factories and destroyed equipment to protest the introduction of mechanical knitting machines, spinning frames, power looms, and other new technologies. Their supposed leader, Ned Ludd, probably was fictitious, but the word *Luddite* is still in use today to refer to a person who opposes technological change.

The Third Industrial Revolution

The advent of the Information Age, around 1970, is usually considered the third industrial revolution. The full impact of this revolution is yet to be felt, as we are still on the thick of it. Whether there will be a fourth industrial revolution that can be clearly separated from the third, remains an open question. Suggesting answers to this question is, in fact, one of the objectives of this book.

Of particular interest for present purposes are the technological developments related to information processing. I believe that, ultimately, information-processing technologies will outpace almost all other existing technologies, and that in the not-so-distant future they will supersede those technologies. The reason for this is that information processing and the ability to record, store, and transmit knowledge are at the origin of most human activities, and will become progressively more important.

The first human need to process information probably arose in the context of the need to keep accurate data about stored supplies and agricultural production. For a hunter-gatherer, there was little need to write down and pass information to others or to future generations. Writing became necessary when complex societies evolved and it became necessary to write down how many sheep should be paid in taxes or how much land someone owned. It also became necessary to store and transmit the elaborate social codes that were required to keep complex societies working.

Creating a written language is so complex a task that, unlike many other technologies, it has probably evolved independently only a few times. To invent a written language, one must figure out how to decompose a sentence or idea into small units, agree on a unified system with which to write down these units, and, to realize its full value, make sure that the whole thing can be understood by third parties. Although other independently developed writing systems may have appeared (Chinese and Egyptian systems among them), the only commonly agreed upon independent developments of writing took place in Mesopotamia between 3500 BC and 3000 BC and in America around 600 BC (Gaur 1992). It is still an open question whether writing systems developed in Egypt around 3200 BC and in China around 1200 BC were independent or whether they were derived from the Mesopotamia cuneiform script. Many other writing systems were developed by borrowing concepts developed by the inventors of these original scripts.

It is now believed that the first writing systems were developed to keep track of amounts of commodities due or produced. Later, writing evolved to be able to register general words in the language and thus to be used to record tales, histories, and even laws. A famous early set of laws is the

Code of Hammurabi from ancient Mesopotamia, which dates from the seventeenth century BC. One of the oldest and most complex writings ever deciphered, the code was enacted by Hammurabi, king of Babylonia. Partial copies exist on a stone stele and in various clay tablets. The code itself consists of hundreds of laws that describe the appropriate actions to be taken when specific rules are violated by people of different social statuses.

Writing was a crucial development in information processing because, for the first time, it was possible to transmit knowledge at a distance and, more crucially, over time spans that transcended memory and even the lifetime of the writer. For the first time in history, a piece of information could be preserved, improved by others, and easily copied and distributed.

The associated ability to count, record, and process numerical quantities was at the origin of writing and developed in parallel with the written language, leading to the fundamentally important development of mathematics. The earliest known application of mathematics arose in response to practical needs in agriculture, business, and industry. In Egypt and in Mesopotamia, in the second and third millennia BC, math was used to survey and measure quantities. Similar developments took place in India, in China, and elsewhere. Early mathematics had an empirical nature and was based on simple arithmetic.

The earliest recorded significant development took place when the Greeks recognized the need for an axiomatic approach to mathematics and developed geometry, trigonometry, deductive systems, and arithmetic. Many famous Greeks contributed to the development of mathematics. Thales, Pythagoras, Plato, Aristotle, Hippocrates, and Euclid were fundamental to the development of many concepts familiar to us today. The Chinese and the Arabs took up mathematics where the Greeks left it, and came up with many important developments. In addition, the Arabs preserved the work of the Greeks, which was then translated and augmented. In what is now Baghdad, Al-Khowarizmi, one of the major mathematicians of his time, introduced Hindu-Arabic numerals and concepts of algebra. The word *algorithm* was derived from his name.

Further developments in mathematics are so numerous and complex that they can't be described properly here, even briefly. Although it took thousands of years for mathematics to progress from simple arithmetic concepts to the ideas of geometry, algebra, and trigonometry developed by the

Greeks, it took less than 500 years to develop the phenomenal edifice of modern mathematics.

One particular aspect of mathematics that deserves special mention here is the theory of computation. Computation is the process by which some calculation is performed in accordance with a well-defined model described by a sequence of operations. Computation can be performed using analog or digital devices. In analog computation, some physical quantity (e.g., displacement, weight, or volume of liquid) is used to model the phenomena under study. In digital computation, discrete representations of the phenomena under study, represented by numbers or symbols, are manipulated in order to yield the desired results.

A number of devices that perform analog computation have been developed over the centuries. One of the most remarkable—the Antikythera mechanism, which has been dated to somewhere between 150 and 100 BC—is a complex analog computer of Greek origin that uses a complex set of interlocked gears to compute the positions of the sun, the moon, and perhaps other celestial bodies (Marchant 2008). Analog computation has a long tradition and includes the astrolabe, attributed to Hipparchus (c. 190–120 BC), and the slide rule, invented by William Oughtred in the seventeenth century and based on John Napier's concept of logarithms. A highly useful and effective tool, the slide rule has been used by many engineers still alive today, perhaps even by some readers of this book.

The twentieth century saw many designs and applications of analog computers. Two examples are the Mark I Fire Control Computer, which was installed on many US Navy ships, and the MONIAC, a hydraulic computer that was created in 1949 by William Phillips to model the economy of the United Kingdom (Bissell 2007; Phillips 1950). Analog computers, however, have many limitations in their flexibility, mainly because each analog computer is conceived and built for one specific application. General-purpose analog computers are conceptually possible (and the slide rule is a good example), but in general analog computers are designed to perform specific tasks and cannot be used for other tasks.

Digital computers, on the other hand, are universal machines. By simply changing the program such a computer is executing, one can get it to perform a variety of tasks. Although sophisticated tools to help with arithmetic operations (such as the abacus, developed around 500 BC) have existed for thousands of years, the idea of completely automatic computation didn't appear until much more recently. Thomas Hobbes—probably the first to present a clearly mechanistic view of the workings of the human brain—clearly thought of the brain as nothing more than a computer:

> When a man reasoneth, he does nothing else but conceive a sum total, from addition of parcels; or conceive a remainder, from subtraction of one sum from another… These operations are not incident to numbers only, but to all manner of things that can be added together, and taken one out of another. (Hobbes 1651)

However, humankind had to wait until the nineteenth century for Charles Babbage to create the first design of a general-purpose digital computer (the Analytical Engine), and for his contemporary Ada Lovelace to take the crucial step of understanding that a digital computer could do much more than merely "crunch numbers."

The Surprising Properties of Exponential Trends

It is common to almost all areas of technology that new technologies are introduced at an ever-increasing rate. For thousands of years progress is slow, and for many generations people live pretty much as their parents and grandparents did; then some new technology is introduced, and a much shorter time span elapses before the next technological development takes place. Reaching a new threshold in a fraction of the time spent to reach the previous threshold is characteristic of a particular mathematical function called the *exponential*.

In mathematics, an exponential function is a function of the form a^n. Because $a^{n+1} = a \times a^n$, the value of the function at the point $n + 1$ is equal to the value of the function at point n multiplied by the basis of the exponential, a. For $n = 0$, the function takes the value 1, since raising any number (other than zero) to the power 0 gives the value 1.

Of interest here is the case in which a, the basis, is greater than 1. It may be much greater or only slightly greater. In the former case, the exponential will grow very rapidly, and even for small n the function will quickly reach very large values. For instance, the function 10^n will grow very rapidly even for small n. This is intuitive, and it shouldn't surprise us. What is less intuitive is that a function of the form a^n will grow very rapidly even when the base is a small number, such as 2, or even a smaller number, such as 1.1.

I will discuss two types of exponential growth, both of them relevant to the discussions that lie ahead. The first type is connected with the evolution of a function that grows exponentially with time. The second is the exponential growth associated with the combinatorial explosion that derives from the combination of simple possibilities that are mutually independent.

The first example I will use to illustrate the surprising properties of exponential trends will be the well-known history of the inventor of chess and the emperor of China. Legend has it that the emperor became very fond of a game that had just been invented—chess. He liked the game so much that he summoned the inventor to the imperial court and told him that he, the emperor of the most powerful country in the world, would grant him any request. The inventor of chess, a clever but poor man, knew that the emperor valued humility in others but didn't practice it himself. "Your Imperial Majesty," he said, "I am a humble person and I only ask for a small compensation to help me feed my family. If you are so kind as to agree with this request, I ask only that you reward me with one grain of rice for the first square of the chessboard, two grains for the second, four grains for the third square, and so on, doubling in each square until we reach the last square of the board." The emperor, surprised by the seemingly modest request, asked his servants to fetch the necessary amount of rice. Only when they tried to compute the amount of rice necessary did it become clear that the entire empire didn't generate enough rice.

In this case, one is faced with an exponential function with base 2. If the number of grains of rice doubles for every square in relation to the previous square, the total number of grains will be

$$2^0 + 2^1 + 2^2 + 2^3 + \ldots + 2^{63} = 2^{64} - 1,$$

since there are 64 squares on the chessboard. This is approximately equal to 2×10^{19} grains of rice, or, roughly, 4×10^{14} kilograms of rice, since there are roughly 50,000 grains of rice per kilogram. This is more than 500 times the yearly production of rice in today's world, which is approximately 600 million tons, or 6×10^{11} kilograms. It is no surprise the emperor could not grant the request of the inventor, nor is it surprising he was deceived into thinking the request was modest.

The second example is from Charles Darwin's paradigm-changing book *On the Origin of the Species by Means of Natural Selection* (1859). In the following passage from that work, Darwin presents the argument that the exponential growth inherent to animal reproduction must be controlled by selective pressures, because otherwise the descendants of a single species would occupy the entire planet:

> There is no exception to the rule that every organic being naturally increase at so high a rate that if not destroyed, the Earth would soon be covered by the progeny of a single pair. ... The Elephant is reckoned to be the slowest breeder of all known animals, and I have taken some pains to estimate its probable minimum rate of natural increase: it will be under the mark to assume that it breeds when thirty years old, and goes on breeding till ninety years old, bringing forth three pairs of young in this interval; if this be so, at the end of the fifth century there would be alive fifteen million elephants, descended from the first pair.

Even though Darwin got the numbers wrong (as Lord Kelvin soon pointed out), he got the idea right. If one plugs in the numbers, it is easy to verify there will be only about 13,000 elephants alive after 500 years. If one calls a period of 30 years one generation, the number of elephants alive in generation n is given by $a_n = 2 \times a_{n-1} - a_{n-3}$. This equation implies that the ratio of the number of elephants alive in each generation to the number in the previous generation rapidly converges to the golden ratio, 1.618.

The surprising properties of exponential growth end up vindicating the essence of Darwin's argument. In fact, even though only 14 elephants are alive after 100 years, there would be more than 30 million elephants alive after 1,000 years, and after 10,000 years (an instant in time, by evolutionary standards) the number of live elephants would be 1.5×10^{70}. This number compares well with the total number of particles in the universe, estimated to be (with great uncertainty) between 10^{72} and 10^{87}. In this case, the underlying exponential function a^n has a base a equal to approximately 1.618 (if one measures n in generations) or 1.016 (if one measures n in years). Still, over a long period of time, the growth surprises everyone but the most prepared reader.

A somewhat different example of the strange properties of exponential behavior comes from Jorge Luís Borges' short story "The Library of Babel." Borges imagines a library that contains an unthinkably large number of books. Each book has 410 pages, each page 40 lines, and each line 80 letters. The Library contains all the books of this size that can be written with the 25 letters of the particular alphabet used. Any book of this size that one can imagine is in the library. There exists, necessarily, a book with the detailed story of your life, from the time you were born until the day you will die, and there are innumerable translations of this story in all existing languages and in an unimaginable number of non-existing but perfectly coherent languages. There is a book containing everything you said between your birth and some day in your life (at which the book ends, for lack of space); another book picks up where that book left, and so on; another contains what will be your last words. Regrettably, these books are very hard to find, not only because the large majority of books are gibberish but also because you would have no way of telling these books apart from very similar books that are entirely true from the day you were born until some specific day, and totally different from then on.

The mind-boggling concept of the Library of Babel only loses some of its power to confound you when you realize that a library of all the books that can ever be written (of a given size) could never exist, nor could any approximation to it ever be built. The number of books involved is so large as to defy any comparison. The size of the observable universe, measured in cubic millimeters, (about 10^{90} cubic millimeters) is no match even for the number of different lines of 80 characters that can exist (about 10^{112}), much less for the $10^{1,834,097}$ books in the Library of Babel.

Even the astoundingly large numbers involved in the Library of Babel pale in comparison with the immense number of combinations that can be encoded in the DNA of organisms. We now know that the characteristics of human beings, and those of all other known living things, are passed down the generations encoded in the DNA present in each and every cell. The human genome consists of two copies of 23 chromosomes, in a total of approximately 2 × 3 billion DNA bases (2 × 3,036,303,846 for a woman, and a bit less for a man because the Y chromosome is much smaller than the X chromosome). Significantly more than 99 percent of the DNA bases in the human genome are exactly the same in all individuals. The remaining bases code for all the variability present within the species. The majority of the differences between individual human genomes are Single Nucleotide Polymorphisms (SNPs, often pronounced "snips")—locations in the DNA where different people often have different values for a specific DNA base. Since there are two copies of each chromosome (except for the X and Y chromosomes), three different combinations are possible. For instance, in a SNP in which both the base T and the base G occur, some people may have a T/T pair, others a T/G pair, and yet others a G/G pair. SNPs in which more than two bases are present are relatively rare and can be ignored for the purposes of our rough analysis.

Although the exact number of SNPs in the human genome is not known, it is believed to be a few tens of millions (McVean et al. 2012). One of the first projects to sequence the human genome identified about 2 million SNPs (Venter et al. 2001). Assuming that this number is a conservative estimate and that all genomic variation is due to SNPs, we can estimate the total number of possible combinations of SNPs, and therefore of different human genomes, that may exist. We can ignore the slight complication arising from the fact that each one of us has two copies of each chromosome, and consider only that, for each SNP, each human may have one of three possible values. We then obtain the value of $3^{2,000,000}$ different possible arrangements of SNPs in a human genome. This means that, owing to variations in SNPs alone, the number of different humans that could exist is so large as to dwarf even the unimaginably large number of books in the Library of Babel.

Of course, the potential combinations are much more numerous if one considers all the possible genetic combinations of bases in a genome of such a size, and not only the SNPs that exist in humans. Think of the space of $4^{3,000,000,000}$ possible combinations of 3 billion DNA bases, and imagine that, somewhere, lost in the immense universe of non-working genomes, there are genomes encoding for all sorts of fabulous creatures—creatures that have never existed and will never be created or even imagined. Somewhere, among the innumerable combinations of bases that don't correspond to viable organisms, there are an indescribably large number of genetic encodings for all sorts of beings we cannot even begin to imagine—flying humans, unicorns, super-intelligent snake-like creatures, and so on.

Here, for the sake of simplicity, let us consider only the possible combinations of four bases in each copy of the genome, since the extent of this genomic space is so large and so devoid of meaning that any more precise computation doesn't make sense. In particular, diploid organisms, such as humans, have two copies of each chromosome, and therefore one may argue that the size of the space would be closer to $4^{6,000,000,000}$. We will never know what is possible in this design space, since we do not have and never will have the tools that would be needed to explore this formidable universe of possibilities. With luck, we will skim the surface of this infinitely deep sea when the tools of synthetic biology, now being developed, come of age. Daniel Dennett has called this imaginary design space the Library of

Mendel, and a huge library it is—so large that makes the Library of Babel look tiny by comparison (Dennett 1995).

I hope that by now you have been convinced that exponential functions, even those that appear to grow very slowly at first, rapidly reach values that defy our intuition. It turns out that the growth of almost every aspect of technology is well described by a curve that is intrinsically exponential. This exponential growth has been well documented in some specific areas. In 1946, R. Buckminster Fuller published a diagram titled "Profile of the Industrial Revolution as exposed by the chronological rate of acquisition of the basic inventory of cosmic absolutes—the 92 Elements"; it showed the exponential nature of the process clearly (Fuller and McHale 1967). In his 1999 book *The Age of Spiritual Machines*, Ray Kurzweil proposed a "law of accelerating returns" stating that the rate of change in a wide variety of evolutionary systems, including many technologies, tends to increase exponentially. A better-known and more quantitative demonstration of this exponential process is Moore's Law, which states that the number of transistors that can be fitted on a silicon chip doubles periodically (a phenomenon I will discuss in more detail in the next chapter).

It is not reasonable to expect the exponential growth that characterizes many technological developments to have been constant throughout the history of humanity. Significant technological changes, such as the agricultural revolution and the industrial revolution, probably have changed the basis of the exponential—in almost all cases, if not all cases, in the direction of faster growth.

The exponential growth that is a characteristic of many technologies is what leads us to systematically underestimate the state of the art of technology in the near to midterm future. As we will see in the following chapters, many technologies are progressing at an exponential pace. In the next few decades, this exponential pace will change the world so profoundly as to make us truly unable to predict the way of life that will be in place when our grandchildren reach our current age.

Digital technologies will play a central role in the changes that will occur. As looms, tractors, engines, and robots changed the way work was done in the fields and in the factories, digital technology will continue to change the way we perform almost all tasks in our daily lives. With time, digital technologies will make many other technologies less relevant or even obsolete, in the same way they have already made typewriters and telegrams things of the past. Many professions and jobs will also become less necessary or less numerous, as has already happened with typists, bank tellers, and newspaper boys.

This book is an attempt to make an educated guess at what these developments will look like and how they will affect our society, our economy, and, ultimately, our humanity.

A Generation Born with the Computer Age

I was born with the computer age, and I belong to the first generation to design, build, program, use, and understand computers. Yet even I, who should be able to understand the fads and trends brought upon us by the ever-increasing pace of technological change, sometimes feel incapable of keeping up with them. Despite this limitation, I attempt to envision the possible developments of technology and, in particular, the developments in the convergence of computer and biomedical technologies.

It is now clear that computers, or what will come after them, will not only replace all the information delivery devices that exist today, including newspapers, television, radio, and telephones, but will also change deeply the ways we address such basic needs as transportation, food, clothing, and housing. As we come to understand better the way living beings work, we will become able to replace increasingly more complex biological systems with artificial or synthetic substitutes. Our ability to change our environment and, ultimately, our bodies, will profoundly affect the way we live our lives. Already we see signs of this deep change in the ways young people live and interact. We may miss the old times, when people talked instead of texting or walked in the park instead of browsing the Web, but, like it or not, these changes are just preludes to things to come. This may not be as bad as it seems. A common person in the twenty-second century will probably have more freedom, more choices, and more ability to create new things than anyone alive today. He or she will have access to knowledge and to technologies of which we don't even dream. The way someone in the twenty-second century will go about his or her daily life will be, however, as alien to us as today's ways would be to someone from the early nineteenth century.

This book is an attempt to give you a peek at things to come. I cannot predict what technology will bring in 100 years, and I don't think anyone can. But I can try to extrapolate from existing technologies in order to predict how some things may turn out 100 years from now.

For the convenience of the reader, I will not jump forward 100 years in a single step. I will begin by considering the recent and not-so-recent history of technological developments, the trends, and the current state of the technology in a number of critical areas; I will then extrapolate them to the near future. What can reasonably be predicted to be achievable in the near future will give us the boldness required to guess what wonders the ever-increasing speed of technological development will bring. The future will certainly be different from how we may guess it will be, and even more unfamiliar and alien than anything we can predict today. We are more likely to be wrong for being too conservative than for being too bold.

I will begin with the technologies that led to the current digital revolution—a revolution that began with the discovery of electricity and gained pace with the invention of a seemingly humble piece of technology: the transistor, probably the most revolutionary technology ever developed by mankind.

3 From Maxwell to the Internet

A number of important technologies developed rapidly in the nineteenth and twentieth centuries, including electricity, electronics, and computers, but also biotechnology, nanotechnologies, and technologies derived from them. In fact, it is the exponential rate of development of these technologies, more than anything else, that has resulted in the scientific and economic developments of recent decades. In those technologies and in others, that rate is likely to increase.

Four Equations That Changed the World

Phenomena related to electricity—for example, magnetism, lightning, and electrical discharges in fish—remained scientific curiosities until the seventeenth century. Although other scientists had studied electrical phenomena before him, the first extensive and systematic research on electricity was done by Benjamin Franklin in the eighteenth century. In one experiment, reputedly conducted in 1752, Franklin used a kite to capture the electrical energy from a storm.

After the invention of the battery (by Alessandro Volta, in 1800), it was practicable to conduct a variety of experiments with electrical currents. Hans Christian Ørsted and André-Marie Ampere discovered various aspects of the relationship between electric currents and magnetic fields, but it was Michael Faraday's results that made it possible to understand and harness electricity as a useful technology. Faraday performed extensive research on the magnetic fields that appear around an electrical current and established the basis for the concept of the electromagnetic field. He became familiar with basic elements of electrical circuits, such as resistors, capacitors, and inductors, and investigated how electrical circuits work. He also invented the first electrical motors and generators, and it was largely through his efforts that electrical technology came into practical use in the nineteenth century. Faraday envisioned some sort of field (he called it an electrotonic state) surrounding electrical and magnetic devices and imagined that electromagnetic phenomena are caused by changes in it.

However, Faraday had little formal mathematical training. It fell to James Clerk Maxwell to formulate the equations that control the behavior of electromagnetic fields—equations that now bear his name. In 1864 and 1865, Maxwell (who was aware of the results Faraday had obtained) published his results suggesting that electric and magnetic fields were the bases of electricity and magnetism and that they could move through space in waves. Furthermore, he suggested that light itself is an electromagnetic wave, since it propagates at the same speed as electromagnetic fields. The mathematical formulation of his results resulted in a set of equations that are among the most famous and influential in the history of science (Maxwell 1865, 1873).

In its original form, presented in *A Treatise in Electricity and Magnetism*, Maxwell's formulation involved twenty different equations. (Maxwell lacked the tools of modern mathematics, which enable us to describe complex mathematical operations in simple form.) At the time, only a few physicists and engineers understood the full meaning of Maxwell's equations. They were met with great skepticism by many members of the scientific community, among them William Thomson (later known as Lord Kelvin). A number of physicists became heavily involved in understanding and developing Maxwell's work. The historian Bruce Hunt dubbed them the Maxwellians in a book of the same name (Hunt 1991).

One person who soon understood the potential importance of Maxwell's work was Oliver Heaviside. Heaviside dedicated a significant part of his life to the task of reformulating Maxwell's equations and eventually arrived at the four equations now familiar to physicists and engineers. You can see them on T shirts and on the bumpers of cars, even though most people don't recognize them or have forgotten what they mean. On Telegraph Avenue in Berkeley, at the Massachusetts Institute of Technology, and in many other places one can buy a T shirt bearing the image shown in figure 3.1. Such shirts somehow echo Ludwig Boltzmann's question "War es ein Gott der diese Zeichen schrieb?" ("Was it a god who wrote these signs?"), referring to Maxwell's equations while quoting Goethe's *Faust*.

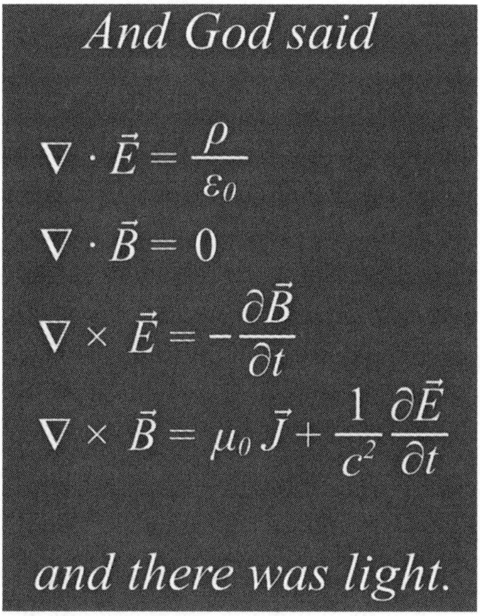

And God said

$$\nabla \cdot \vec{E} = \frac{\rho}{\varepsilon_0}$$

$$\nabla \cdot \vec{B} = 0$$

$$\nabla \times \vec{E} = -\frac{\partial \vec{B}}{\partial t}$$

$$\nabla \times \vec{B} = \mu_0 \vec{J} + \frac{1}{c^2}\frac{\partial \vec{E}}{\partial t}$$

and there was light.

Figure 3.1
The origin of the world according to Oliver Heaviside's reformulation of Maxwell's equations and a popular T shirt.

Hidden in the elegant, and somewhat impenetrable, mathematical formalism of Maxwell's equations are the laws that control the behavior of the electromagnetic fields that are present in almost every electric or electronic device we use today. Maxwell's equations describe how the electric field (E) and the magnetic field (B) interact, and how they are related to other physical entities, including charge density (ρ) and current density (J). The parameters ε_0 and μ_0 are physical constants called the permittivity and permeability of free space, respectively, and they are related by the equation

$$\mu_0 \varepsilon_0 = 1 / c^2,$$

where c is the speed of light.

A field is a mathematical construction that has a specific value for each point in space. This value may be a number, in the case of a scalar field, or a vector, with a direction and intensity, in the case of a vector field. The electric and magnetic fields and the gravitational field are vector fields. At each point in space, they are defined by their direction and their amplitude. One reason Maxwell's work was difficult for his contemporaries to understand was that it is was hard for them to visualize or understand fields, and even harder to understand how field waves could propagate in empty space.

The symbol ∇ represents a mathematical operator that has two different meanings. When applied to a field with the intervening operator $\cdot$, it represents the divergence of the field. The divergence of a field represents the volume density of the outward flux of a vector field from an arbitrarily small volume centered on a certain point. When there is positive divergence in a point, an outward flow of field is created there. The divergence of the gravitational field is mass; the divergence of the electrical field is charge. Charges can be positive or negative, but there is no negative mass (or none was ever found). Positive charges at one point create an outgoing field; negative charges create an incoming field.

When ∇ is applied to a field with the operator $\times$, it represents the curl of the field, which corresponds to an indication of the way the field curls at a specific point. If you imagine a very small ball inside a field that represents the movement of a fluid, such as water, the curl of a field can be visualized as the vector that characterizes the rotating movement of the small ball, because the fluid passes by at slight different speeds on the different sides. The curl is represented by a vector aligned with the axis of rotation of the small ball and has a length proportional to the rotation speed.

The first of Maxwell's equations is equivalent to the statement that the total amount of electric field leaving a volume equals the total sum of charge inside the volume. This is the mathematical formulation of the fact that electric fields are created by charged particles.

The second equation states that the total amount of magnetic field leaving a volume must equal the magnetic field entering the volume. In this respect, the electric field and the magnetic field differ because, whereas there are electrically charged particles that create electric fields, there are no magnetic monopoles, which would create magnetic fields. Magnetic fields are, therefore, always closed loops, because the amount of field entering a volume must equal the amount of field leaving it.

The third equation states that the difference in electrical potential accumulated around a closed loop, which translates into a voltage difference, is equal to the change in time of the magnetic flux through the area enclosed by the loop.

The fourth equation states that electric currents and changes in the electric field flowing through an area are proportional to the magnetic field that circulates around the area.

The understanding of the relationship between electric fields and magnetic fields led, in time, to the development of electricity as the most useful technology of the twentieth century. Heinrich Hertz, in 1888, at what is now Karlsruhe University, was the first to demonstrate experimentally the existence of the electromagnetic waves Maxwell had predicted, and that they could be used to transmit information over a distance.

We now depend on electrical motors and generators to power appliances, to produce the consumer goods we buy, and to process and preserve most of the foods we eat. Televisions, telephones, and computers depend on electric and magnetic fields to send, receive, and store information, and many medical imaging technologies are based on aspects of Maxwell's equations.

Electrical engineers and physicists use Maxwell's equations every day, although sometimes in different or simplified forms. It is interesting to understand, in a simple case, how Maxwell's equations define the behavior of electrical circuits. Let us consider the third equation, which states that the total sum of voltages measured around a closed circuit is zero when there is no change in the magnetic field crossing the area surrounded by the circuit. In the electrical circuit illustrated in figure 3.2, which includes one battery as a voltage source, one capacitor, and one resistor, the application of the third equation leads directly to

$$V_R + V_K - V_C = 0.$$

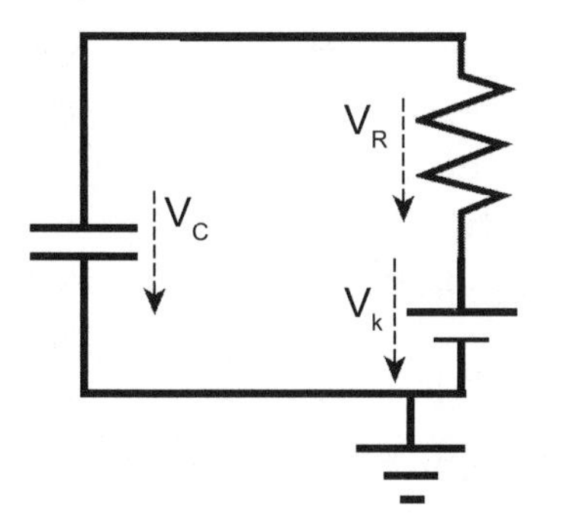

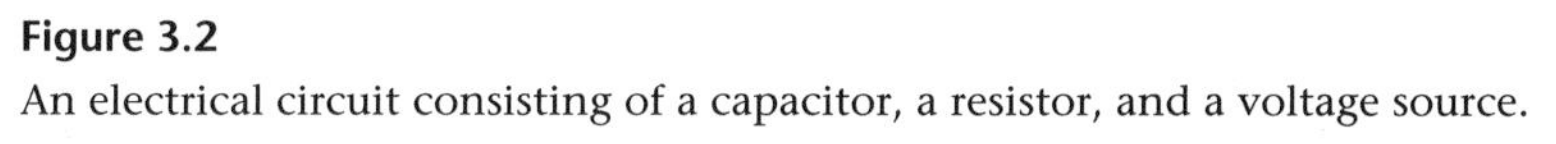

Figure 3.2
An electrical circuit consisting of a capacitor, a resistor, and a voltage source.

This expression results from computing the sum of the voltage drops around the circuit, in a clockwise circulation. Voltage drop V_C is added with a negative sign because it is defined against the direction of the circulation; voltages V_K (in the battery) and V_R (in the resistor) are defined in the direction of the circulation. It turns out that the current through the capacitor, I_C, is the same (and in the opposite direction) as the current through the resistor, I_R, because in this circuit all current flows through the wires.

Maxwell's fourth equation implies that the current through the capacitor is proportional to the variation in time of the electric field between the capacitor plates, leading to the expression $I_C = C\dot{V}_C$, since the electric field ($\vec{E}$, in this case) is proportional to the voltage difference between the capacitor plates, V_C. In this equation, the dot above V_C represents the variation in time of this quantity. Mathematically, it is called the derivative with respect to time, also represented as dV_C/dt.

The linear relation between the current I_R and the voltage V_R across a resistor was described in 1827 by another German physicist, Georg Ohm. Ohm's Law states that there is a linear relation $V_R = RI_R$ between the two variables, given by the value of the resistance R. Putting together these expressions, we obtain

$$-C\dot{V}_C = (V_C - V_K) / R.$$

This expression, which is a direct result of the application of Maxwell's equations and Ohm's Law, is a differential equation with a single unknown, V_C (the voltage across the capacitor, which varies with time). All other parameters in this expression are known, fixed physical quantities. Differential equations of this type relate variables and their variations with respect to time. This particular differential equation can be solved, either by analytical or numerical methods, to yield the value of the voltage across the capacitor as it changes with time. Similar analyses can be used to derive the behaviors of much more complex circuits with thousands or even millions of electrical elements.

We know now that neurons in the brain use the very same electromagnetic fields described by Maxwell's equations to perform their magic. (This will be discussed in chapter 8.) Nowadays, very fast computers are used to perform brain simulation and electrical-circuit simulation by solving Maxwell's equations.

The Century of Physics

The twentieth century has been called the century of physics. Although knowledge of electromagnetism developed rapidly in the nineteenth century, and significant insights about gravity were in place before 1900, physics was the driving force of the major advances in technology that shaped the twentieth century. A comprehensive view of the relationship between matter and energy was first developed in that century, when the strong and weak nuclear forces joined the electromagnetic and gravitational forces to constitute what we see now as the complete set of four interactions that govern the universe.

Albert Einstein's "annus mirabilis papers" of 1905 started the century on a positive note. Einstein's seminal contributions on special relativity, matter, and energy equivalence (1905c), on the photoelectric effect (1905b), and on Brownian motion (1905a) changed physics. Those publications, together with our ever-growing understanding of electromagnetism, started a series of developments that led to today's computer and communications technologies.

The 1920s brought quantum mechanics, a counter-intuitive theory of light and matter that resulted from the work of many physicists. One of the first contributions came from Louis de Broglie, who asserted that particles can behave as waves and that electromagnetic waves sometimes behave like particles. Other essential contributions were made by Erwin Schrödinger (who established, for the first time, the probabilistic base for quantum mechanics) and by Werner Heisenberg (who established the impossibility of precisely and simultaneously measuring the position and the momentum of a particle). The philosophical questions raised by these revolutionary theories remain open today, even though quantum mechanics has proved to be one of the most solid and one of the most precisely tested physical theories of all time.

After World War II, Julian Schwinger, Richard Feynman, and Sin-Itiro Tomonaga independently proposed techniques that solved numerical difficulties with existing quantum theories, opening the way for the establishment of a robust theory of quantum electrodynamics. Feynman was also a talented storyteller and one of the most influential popular science writers of all time. In making complicated things easy to understand, few books match his *QED: The Strange Theory of Light and Matter* (1985).

High-energy physics led to a range of new discoveries, and a whole zoo of sub-atomic particles enriched the universe of particle physics. The idea that fundamental forces are mediated by particles (photons for the electromagnetic force, mesons for the nuclear forces) was verified experimentally. A number of other exotic particles, including positrons (the antimatter version of the electron), anti-protons, anti-neutrons, pions, kaons, muons, taus, neutrinos, and many others—joined the well-known electrons, neutrons, protons, and photons as constituents of matter and energy. At first these particles were only postulated or were found primarily by the ionized trails left by cosmic rays, but with particle accelerators such as those at CERN and Fermilab they were increasingly produced. And even more exotic particles, including the weakly interacting massive particle (WIMP), the W particle, the Z^0 particle, and the elusive Higgs boson, joined the party, and will keep theoretical physicists busy for many years to come in their quest for a unifying theory of physics.

A number of more or less exotic theories were proposed to try to unify gravity with the other three forces (strong interaction, electromagnetic interaction, and weak interaction), using as few free parameters as is possible, in a great unifying theory. These theories attempt to replace the standard model, which unifies the electromagnetic, strong, and weak interactions and which comprises quantum electrodynamics and quantum chromodynamics. Among the most popular are string theories, according to which elementary particles are viewed as oscillating strings in some higher-dimensional space. For example, the popular M-theory (Witten 1995) requires space-time to have eleven dimensions—hardly a parsimonious solution. So far, no theory has been very successful at predicting observed phenomena without careful tuning of parameters after the fact.

At the time of this writing, the Large Hadron Collider at CERN, in Switzerland, represents the latest effort to understand the world of high-energy physics. In 2012, scientists working at CERN were able to detect the elusive Higgs boson, the only particle predicted by the standard model that had never been observed until then. The fact that such a particle had never been observed before is made even more curious by the fact that it should be responsible for the different characteristics of photons (which mediate electromagnetic force, and are massless) and the massive W and Z bosons (which mediate the weak force).

The existing physical theories are highly non-intuitive, even for experts. One of the basic tenets of quantum physics is that the actual outcome of an observation cannot be predicted, although its probability can be computed. This simple fact leads to a number of highly counter-intuitive results, such as the well-known paradox of Schrödinger's cat (Schrödinger 1935) and the Einstein-Podolsky-Rosen paradox (Einstein, Podolsky, and Rosen 1935).

Schrödinger proposed a thought experiment in which the life of a cat that has been placed in a sealed box depends on the state of a radioactive atom that controls, through some mechanism, the release of a poisonous substance. Schrödinger proposed that, until an observation by an external observer took place, the cat would be simultaneously alive and dead, in a quantum superposition of macroscopic states linked to a random subatomic event with a probability of occurrence that depends on the laws of quantum mechanics.

Schrödinger's thought experiment, conceived to illustrate the paradoxes that result from the standard interpretation of quantum theory, was, in part, a response to the Einstein-Podolsky-Rosen (EPR) paradox, in which the authors imagine two particles that are entangled in their quantum states, creating a situation such that measuring a characteristic of one particle instantaneously makes a related characteristic take a specific value for the other particle. The conceptual problem arises because pairs of entangled particles are created in many physical experiments, and the particles can travel far from each other after being created. Measuring the state of one of the particles forces the state of the other to take a specific value, even if the other particle is light-years away. For example, if a particle with zero spin, such as a photon, decays into an electron and a positron (as happens in PET imaging, which we will discuss in chapter 9), each of the products of the decay must have a spin, since the spins must be opposite on account of the conservation of spin. But only when one of the particles is measured can its spin be known. At that exact moment, the spin of the other particle will take the opposite value, no matter how far apart they are in space. Such "spooky" instant action at a distance was deemed impossible, according to the theory of relativity, because it could be used to transmit information at a speed exceeding that of light. Posterior interpretations have shown that the entanglement mechanism cannot in fact be used to transmit information, but the "spooky action-at-a-distance" mechanism (Einstein's expression) remains as obscure as ever.

The discussions and excitement that accompanied the aforementioned paradoxes and other paradoxes created by quantum mechanics would probably have gone mostly unnoticed by the general public had it not been for their effects on the real world. Many things we use in our daily lives would not be very different if physics had not developed the way it did. We are still raising cattle and cultivating crops much as our forebears did, and we go around in cars, trains, and planes that, to a first approximation, have resulted from the first industrial revolutions and could have been developed without the advantages of modern physics.

There are, however, two notable exceptions. The first was noticed by the whole world on August 6, 1945, when an atomic bomb with a power equivalent to that of 12–15 kilotons of TNT was detonated over Hiroshima. The secret Manhattan Project, headed by the physicist J. Robert Oppenheimer, had been commissioned, some years before, to explicitly explore the possibility of using Einstein's equation $E = mc^2$ to produce a bomb far more powerful than any that existed before. The further developments of fission-based and fusion-based bombs are well known.

The second exception, less noticeable at first, was the development of an apparently simple device called the transistor. Forty years later, it would launch humanity into the most profound revolution it has ever witnessed.

Transistors, Chips, and Microprocessors

The first patent for a transistor-like device (Lilienfeld 1930) dates from 1925. Experimental work alone led to the discovery of the effect that makes transistors possible. However, the understanding of quantum mechanics and the resulting field of solid-state physics were instrumental in the realization that the electrical properties of semiconductors could be used to obtain behaviors that could not be obtained using simpler electrical components, such as resistors and capacitors.

In 1947, researchers at Bell Telephone Laboratories observed that when electrical contacts were applied to a germanium crystal (a semiconductor) the output signal had more power than the input signal. For this discovery, which may have been the most important invention ever, William Shockley, John Bardeen, and Walter Brattain received the Nobel Prize in Physics in 1956. Shockley, foreseeing that such devices could be used for many important applications, set up the Shockley Semiconductor Laboratory in Mountain View, California. He was at the origin of the dramatic transformations that would eventually lead to the emergence of modern computer technology.

A transistor is a simple device with three terminals, one of them a controlling input. By varying the voltage at this input, a large change in electrical current through the two other terminals of the device can be obtained. This can be used to amplify a sound captured by a microphone or to create a powerful radio wave. A transistor can also be used as a controlled switch.

The first transistors were bipolar junction transistors. In such transistors, a small current also flows through the controlling input, called the base. Other types of transistors eventually came to dominate the technology. The metal-oxide-semiconductor field-effect transistor (MOSFET) is based on different physical principles, but the basic result is the same: A change in voltage in the controlling input (in this case called the *gate*) creates a significant change in current through the two other terminals of the device (in this case called the *source* and the *drain*). From the simple description just given, it may be a little difficult to understand why such a device would lead to the enormous changes that have occurred in society in the last thirty years, and to the even larger changes that will take place in coming decades. Vacuum tubes, first built in the early twentieth century, exhibit a behavior similar to that of transistors, and can indeed be used for many of the same purposes. Even today, vacuum tubes are still used in some audio amplifiers and in other niche applications. Even though transistors are much more reliable and break down much less frequently than vacuum tubes, the critical difference, which took several decades to exploit to its full potential, is that, unlike vacuum tubes, transistors can be made very small, in very large numbers, and at a very small cost per unit. Although the British engineer Geoffrey Dummer was the first to propose the idea that many transistors could be packed into an integrated circuit, Jack Kilby and Robert Noyce deserve the credit for realizing the first such circuits, which required no connecting wires between the transistors.

Transistors, tightly packed into integrated circuits, have many uses. They can be used to amplify, manipulate, and generate analog signals, and indeed many devices, such as radio and television receivers, sound amplifiers, cellular phones, and GPS receivers, use them for that purpose. Transistors have enabled circuit designers to pack into very small volumes amplifiers and other signal-processing elements that could not have been built with vacuum tubes or that, if built with them, would have occupied a lot of space and weighed several tons. The development of personal mobile communications was made possible, in large measure, by this particular application of transistors.

However, the huge effect transistor technology has in our lives is due even more to the fact that integrated circuits with large numbers of transistors can be easily mass produced and to the fact that transistors can be used to process digital information by behaving as controlled on-off switches. Digital computers manipulate data in binary form. Numbers, text, images, sounds, and all other types of information are stored in the form of very long strings of bits (binary digits—zeroes and ones). These binary digits are manipulated by digital circuits and stored in digital memories. Digital circuits and memories are built out of logic gates, all of them made of transistors. For example, a *nand* gate has two inputs and one output. Its output is 0 if and only if both inputs are 1. This gate, one of the simplest gates possible, is built using four transistors, as shown in figure 3.3a, which shows

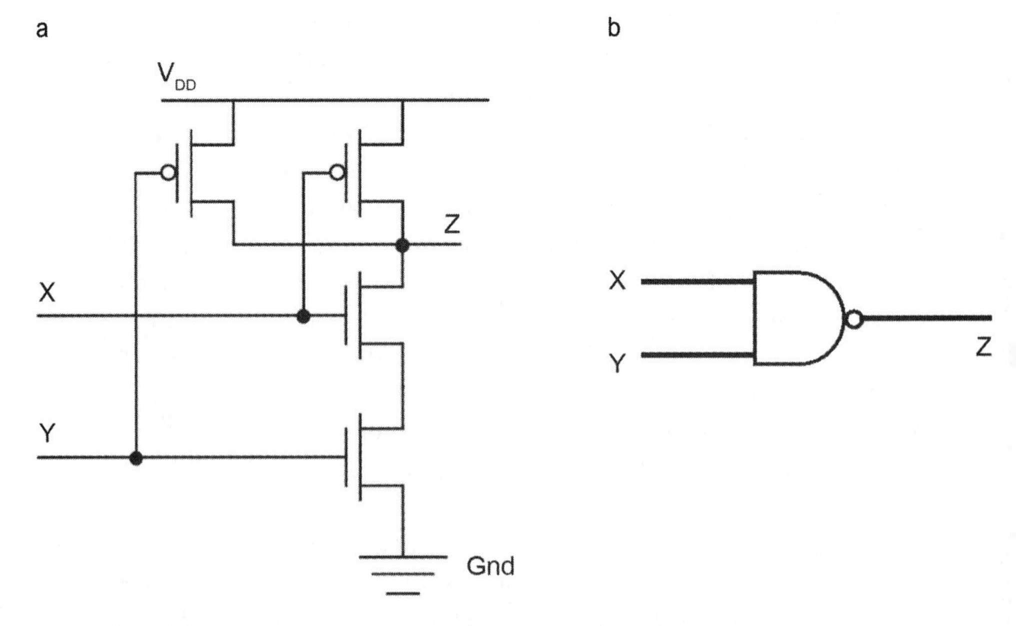

Figure 3.3
(a) A *nand* gate made of MOSFET transistors. (b) The logic symbol for a *nand* gate.

four MOSFETs: two of type N at the bottom and two of type P at the top. A type N MOSFET behaves like a closed switch if the gate is held at a high voltage, and like an open switch if the gate is held at a low voltage. A P MOSFET, which can be recognized because it has a small circle on the gate, behaves in the opposite way. It behaves like a closed switch when the gate is held at a low voltage, and like an open switch when the gate is held at a high voltage.

If both X and Y (the controlling inputs of the transistors) are at logical value 1 (typically the supply voltage, V_{DD}), the two transistors shown at the bottom of figure 3.3a work as closed switches and connect the output Z to the ground, which corresponds to the logical value 0. If either X or Y (or both) is at the logical value 0 (typically ground, or 0 volt), or if both of them are, then at least one of the two transistors at the bottom of figure 3.3a works as an open switch and one (or both) or the top transistors works as a closed switch, pulling the value of Z up to $V_{DD,}$ which corresponds to logical value 1. This corresponds in effect to computing $Z = \overline{X \wedge Y}$, where the bar denotes negation and $\wedge$ denotes the logic operation *and*. This is the function with the so-called truth table shown here as table 3.1.

Besides *nand* gates there are many other types of logic gates, used to compute different logic functions. An *inverter* outputs the opposite logic value of its input and is, in practice, a simplified *nand* gate with only two transistors, one of type P and one of type N. An *and* gate, which computes the conjunction of the logical values on the inputs and outputs 1 only when both inputs are 1, can be obtained by using a *nand* gate followed by an *inverter*. Other types of logic gates, including *or* gates, *exclusive-or* gates, and *nor* gates, can be built using different arrangements of transistors and basic gates. More complex digital circuits are built from these simple logic gates.

Table 3.1
A truth table for logic function *nand*.

X	Y	Z
0	0	1
0	1	1
1	0	1
1	1	0

Nand gates are somewhat special in the sense that any logic function can be built out of *nand* gates alone (Sheffer 1913). In fact, *nand* gates can be combined to compute any logic function or any arithmetic function over binary numbers. This property results from the fact that *nand* gates are *complete,* meaning that they can be used to create any logic function, no matter how complex. For instance, *nand* gates can be used to implement the *two-bit exclusive-or* function, a function that evaluates to 1 when exactly one of the input bits is at 1. They can also be used to implement the *three-bit majority function,* which evaluates to 1 when two or more bits are at 1. Figure 3.4 illustrates how the *two-bit exclusive-or* function (which evaluates to 1 when exactly one input is 1) and the *three-bit majority* function (which evaluates to 1 when at least two inputs are 1) can be implemented using *nand* gates.

In fact, circuits built entirely of *nand* gates can compute additions, subtractions, multiplications, and divisions of numbers written in binary,

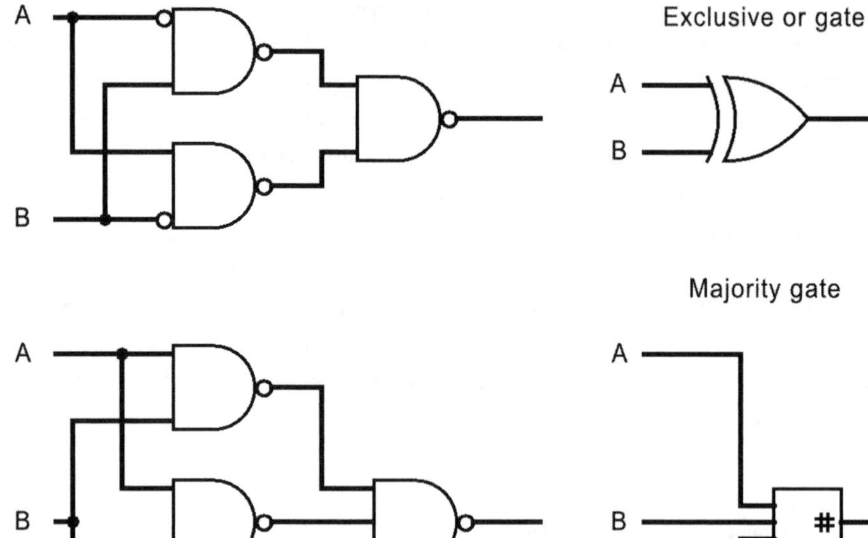

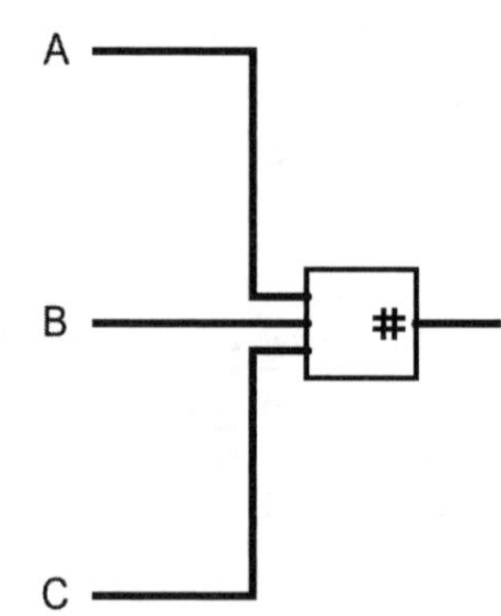

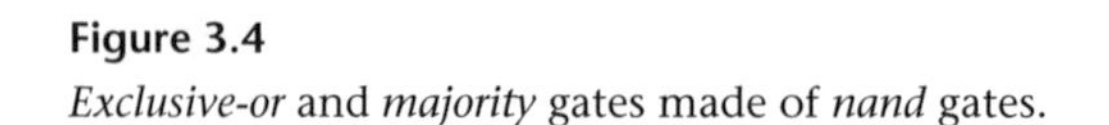

Figure 3.4
Exclusive-or and *majority* gates made of *nand* gates.

as well as any other functions that can be computed by logic circuits. In practice, almost all complex digital circuits are built using *nand* gates, *nor* gates, and *inverters*, because these gates not only compute the logic function but also regenerate the level of the electrical signal, so it can be used again as input to other logic gates. Conceptually, a complete computer can be built of *nand* gates alone; in fact, a number of them have been built in that way.

Internally, computers manipulate only numbers written in binary form. Although we are accustomed to the decimal numbering system, which uses the digits 0 through 9, there is nothing special about base 10. That base probably is used because humans have ten fingers and so it seemed to be natural. A number written in base 10 is actually a compact way to describe a weighted sum of powers of 10. For instance, the number 121 represents

$$1 \times 10^2 + 2 \times 10^1 + 1 \times 10^0,$$

because every position in a number corresponds to a specific power of 10.

When writing numbers in other bases, one replaces the number 10 with the value of the base used. If the base is smaller than 10, fewer than ten symbols are required to represent each digit. The same number, 121 in base 10, when written in base 4, becomes

$$1 \times 4^3 + 3 \times 4^2 + 2 \times 4^1 + 1 \times 4^0$$

(which also can be written as 1321_4, the subscript denoting the base). In base 2, powers of 2 are used and there are only two digits, 0 and 1, which can be conveniently represented by two electrical voltage levels—for example, 0 and 5 V. The same number, 121_{10} in base 10, becomes, in base 2, 1111001_2, which stands for

$$1 \times 2^6 + 1 \times 2^5 + 1 \times 2^4 + 1 \times 2^3 + 0 \times 2^2 + 0 \times 2^1 + 1 \times 2^0.$$

Arithmetic operations between numbers written in base 2 are performed using logic circuits that compute the desired functions. For instance, if one is using four-bit numbers and wishes to add 7_{10} and 2_{10}, then one must add the equivalent representations in base 2, which are 0111_2 and 0010_2.

The addition algorithm, shown in figure 3.5, is the one we all learned in elementary school. The algorithm consists in adding the digits, column by column, starting from the right, and writing the carry bit from the previous column above the next column to the left. The only difference is that for each column there are only four possible combinations of inputs, since

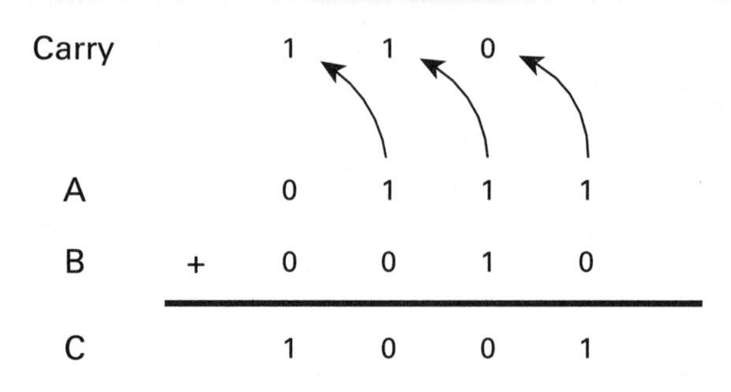

Figure 3.5
Adding together the numbers 0111_2 and 0010_2.

Table 3.2
Logic functions for the addition of two binary digits. *C* gives the value of the result bit; *Cout* gives the result of the carry bit.

A	*B*	*Cin*	*C*	*Cout*
0	0	0	0	0
0	0	1	1	0
0	1	0	1	0
0	1	1	0	1
1	0	0	1	0
1	0	1	0	1
1	1	0	0	1
1	1	1	1	1

each digit can take only the value 0 or 1. However, since the carry in bit can also take two possible values (either there is a carry or there is not), there are a total of eight possible combinations for each column. Those eight combinations are listed in table 3.2, together with the desired values for the output, *C*, and the carry bit, *Cout*, which must be added to the bits in the next column to the left. The carry is simultaneously an output of a column (*Cout*) and an input in the next column to the left (*Cin*).

It is easy to verify by inspection that the function *C* is given by the *exclusive-or* of the three input bits, a function that takes the value 1 when an odd number of bits are 1. The function *Cout* is given by the majority function of the same three input bits.

Therefore, the circuit on the left of figure 3.6 computes the output (*C*) and the carry out (*Cout*) of its inputs, *A*, *B*, and *Cin*. More interestingly, by

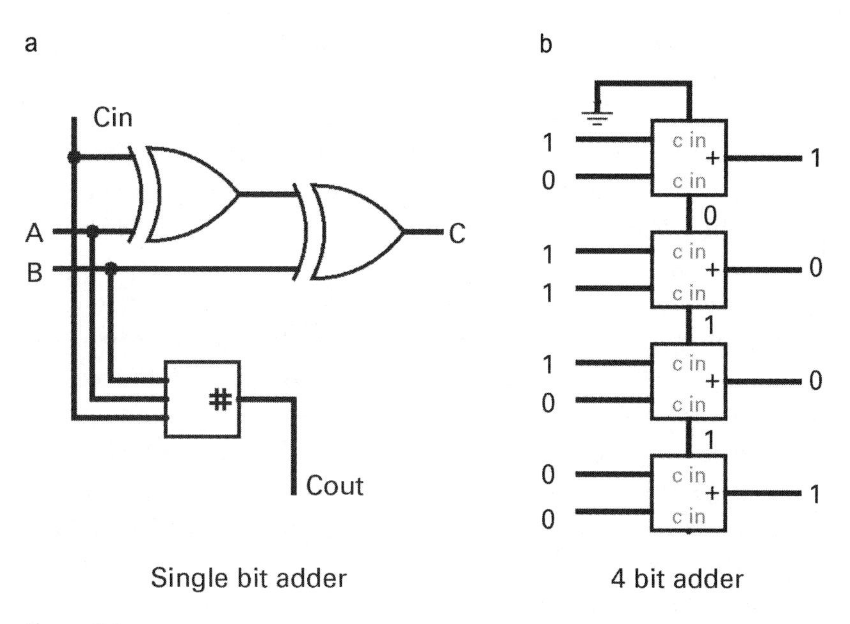

Figure 3.6
(a) A single-bit adder. (b) A four-bit adder, shown adding the numbers 0111 and 0010 in binary..

wiring together four of these circuits one obtains a four-bit adder, like the one shown on the right of figure 3.6. In this four-bit adder, the topmost single-bit adder adds together the least significant bits of numbers A and B and the carry bit propagates through the chain of adders. This circuit performs the addition of two four-bit numbers, using the algorithm (and the values in the example) from figure 3.5.

More complex circuits (for example, multipliers, which compute the product of two binary numbers) can be built out of these basic blocks. Multiplications can be performed by specialized circuits or by adding the same number multiple times, using the same algorithm we learned in elementary school. Building blocks such as adders and multipliers can then be interconnected to form digital circuits that are general in the sense that they can execute any sequence of basic operations between binary numbers. The humble transistor thus became the workhorse of the computer industry, making it possible to build cheaply and effectively the adders and multipliers Thomas Hobbes imagined, in 1651, as the basis of all human reasoning and memory.

Transistors and logic gates, when arranged in circuits that store binary values over long periods of time, can also be used to build computer memories. Such memories, which can store billions or trillions of bits, are part of every computer in use today. Transistors can, therefore, be used to build general-purpose circuits that compute all possible logic operations quickly, cheaply, and effectively. A sufficiently complex digital circuit can be instructed to add the contents of one memory position to the contents of another memory position, and to store the result in a third memory position. Digital circuits flexible enough to perform these and other similar operations are called *Central Processing Units* (CPUs). A CPU is the brain of every computer and almost every advanced electronic device we use today. CPUs execute programs, which are simply long sequences of very simple operations. (In the next chapter, I will explain how CPUs became the brains of modern computers as the result of pioneering work by Alan Turing, John von Neumann, and many, many others.)

The first digital computers were built by interconnecting logic gates made from vacuum tubes. They were bulky, slow, and unreliable. The ENIAC—the first fully electronic digital computer, announced in 1946—contained more than 17,000 vacuum tubes, weighted more than 27 tons, and occupied more than 600 square feet.

When computers based on discrete transistors became the norm, large savings in area occupied and in power consumed were achieved. But the real breakthrough came when designers working for the Intel Corporation recognized that they could use a single chip to implement a CPU. Such chips came to be called *microprocessors*. The first single-chip CPU—the 4004 processor, released in 1971—manipulated four-bit binary numbers, had 2,300 transistors, and weighted less than a gram. A present-day high-end microprocessor has more than 3 billion transistors packed in an area about the size of a postage stamp (Riedlinger et al. 2012).

Nowadays, transistors are mass produced at the rate of roughly 150 trillion (1.5×10^{14}) per second. More than 3×10^{21} of them have been produced to date. This number compares well with some estimates of the total number of grains of sand on Earth. In only a few years, we will have produced more transistors than there are synapses in the brains of all human beings currently alive.

The number of transistors in microprocessors has grown rapidly since 1971, following an approximately exponential curve which is known as

Moore's Law. (In 1965, Intel's co-founder, Gordon Moore, first noticed that the number of transistors that could be placed inexpensively on an integrated circuit increased exponentially over time, doubling approximately every two years.) Figure 3.7 depicts the increase in the number of transistors in Intel's microprocessors since the advent of the 4004. Note that, for convenience, the number of transistors is shown in a logarithmic scale. Although the graph is relative to only a small number of microprocessors from one supplier, it illustrates a typical case of Moore's Law. In this case, the number of transistors in microprocessors has increased by a factor of a little more than 2^{20} in 41 years. This corresponds roughly to a factor of 2 every two years.

Many measures of the evolution of digital electronic devices have obeyed a law similar to Moore's. Processing speed, memory capacity, and sensor sensitivity have all been improving at an exponential rate that approaches the rate predicted by Moore's Law. This exponential increase is at the origin of the impact digital electronics had in nearly every aspect of our lives. In fact, Moore's Law and the related exponential evolution of digital technologies are at the origin of many of the events that have changed society profoundly in recent decades.

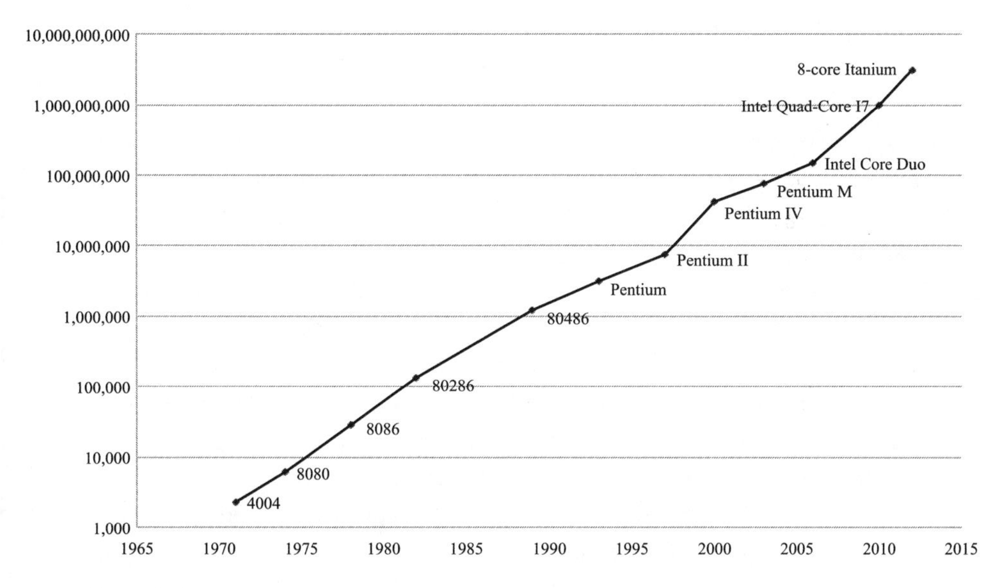

Figure 3.7
Evolution of the number of transistors of Intel microprocessors.

Other digital technologies have also been improving at an exponential rate, though in ways that are somewhat independent of Moore's Law. Kryder's Law states that the number of bits that can be stored in a given area in a magnetic disk approximately doubles every 13 months. Larry Roberts has kept detailed data on the improvements of communication equipment and has observed that the cost per fixed communication capacity has decreased exponentially over a period of more than ten years.

For the case of Moore's Law, the progress is even more dramatic than that shown in figure 3.7, since the speed of processors has also been increasing. In a very simplified view of processor performance, the computational power increases with both the number of transistors and the speed of the processor. Therefore, processors have increased in computational power by a factor of about 30 billion over the period 1971–2012, which corresponds to a doubling of computational power every 14 months.

The technological advances in digital technologies that led to this exponential growth are unparalleled in other fields of science, with a single exception (which I will address in chapter 7): DNA sequencing. The transportation, energy, and building industries have also seen significant advances in recent decades. None of those industries, however, was subject to the type of exponential growth that characterized semiconductor technology. To put things in perspective, consider the fuel efficiency of automobiles. In approximately the same period as was discussed above, the fuel efficiency of passenger cars went from approximately 20 miles per gallon to approximately 35. If cars had experienced the same improvement in efficiency over the last 40 years as computers, the average passenger car would be able to go around the Earth more than a million times on one gallon of fuel.

The exponential pace of progress in integrated circuits has fueled the development of information and communication technologies. Computers, interconnected by high-speed networks made possible by digital circuit technologies, became, in time, the World Wide Web—a gigantic network that interconnects a significant fraction of all the computers in existence.

There is significant evidence that, after 25 years, Moore's Law is running out of steam—that the number of transistors that can be packed onto a chip is not increasing as rapidly as in the past. But it is likely that other technologies will come into play, resulting in a continuous (albeit slower) increase in the power of computers.

The Rise of the Internet

In the early days of digital computers, they were used mostly to replace human computers in scientific and military applications. Before digital computers, scientific and military tables were computed by large teams of people called human computers. In the early 1960s, mainframes (large computers that occupied entire rooms) began to be used in business applications. However, only in recent decades has it become clear that computers are bound to become the most pervasive appliance ever created.

History is replete with greatly understated evaluations of the future developments of computers. A probably apocryphal story has it that in 1943 Thomas Watson, the founder of IBM, suggested that there would be a worldwide market for perhaps five computers. As recently as 1977, Ken Olson, chairman and founder of the Digital Equipment Corporation, was quoted as saying "There is no reason anyone would want a computer in their home." In 1981, Bill Gates, chairman of Microsoft, supposedly stated that 640 kilobytes of memory ought to be enough for anyone. All these predictions vastly underestimated the development of computer technology and the magnitude of its pervasiveness in the modern world.

However, it was not until the advent of the World Wide Web (which began almost unnoticeably in 1989 as a proposal to interlink documents in different computers so that readers of one document could easily access other related documents) that computers entered most people's daily lives. The full power of the idea of the World Wide Web was unleashed by the Internet, a vast network of computers, interconnected by high-speed communications equipment, that spans the world. The Internet was born in the early 1970s when a group of researchers proposed a set of communication protocols known as TCP/IP and created the first experimental networks interconnecting different institutions.

The TCP/IP protocol soon enabled thousands of computers, and later millions, to become interconnected and to exchange files and documents. The World Wide Web was made easily accessible, even to non-expert users, by the development of Web browsers—programs that display documents and can be used to easily follow hypertext links. The first widely used Web browser—Mosaic, developed by a team at the University of Illinois at Urbana-Champaign, led by Marc Andreessen—was released in 1993, and represented a turning point for the World Wide Web. The growth of the

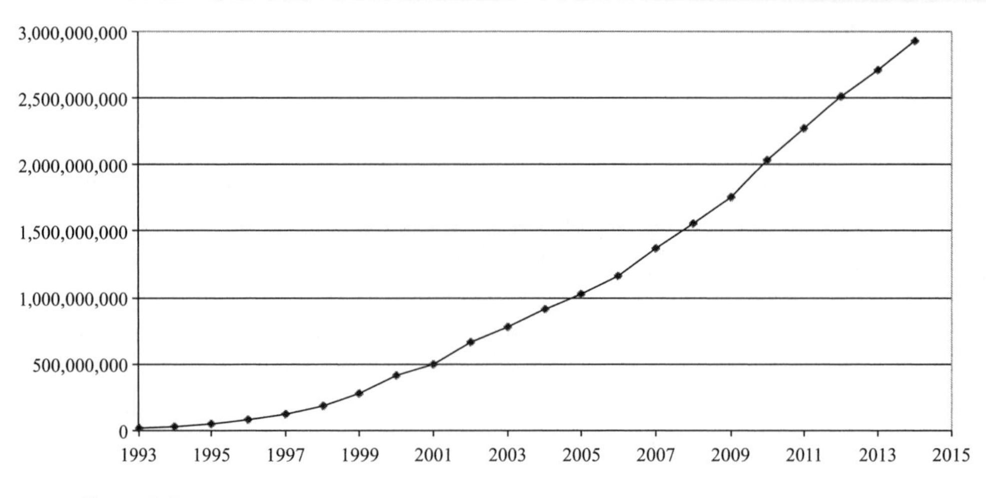

Figure 3.8
Evolution of the number of users of the Internet.

Internet and the popularity of the World Wide Web took the world by surprise. Arguably, no phenomenon has changed so rapidly and so completely the culture and daily life around the world as the Internet.

Figure 3.8 plots the number of users since the beginning of the Internet. The number of users grew from just a few in 1993 to a significant fraction of the world population in twenty years. Probably no other technology (except some that were developed on top of the Web) has changed the world as rapidly as the World Wide Web has.

Initially, the World Wide Web gave users access to documents stored in servers (large computers, maintained by professionals, that serve many users at once). Development of the content stored in these servers was done mostly by professionals or by advanced users. However, with the development of more user-friendly applications and interfaces it became possible for almost any user to create content, be it text, a blog, a picture, or a movie. That led to what is known as *Web 2.0* and to what is known in network theory as quadratic growth of the Web's utility. For example, if the number of users doubles, and all of them contribute to enriching the Web, the amount of knowledge that can be used by the whole community grows by a factor of 4, since twice as many people have access to twice as much knowledge. This quadratic growth of the network utility has fueled development of new applications and uses of the World Wide Web, many of them unexpected only a few years ago.

The Digital Economy

Easy access to the enormous amounts of information available on the World Wide Web, by itself, would have been enough to change the world. Many of us can still remember the effort that was required to find information on a topic specialized enough not to have an entry in a standard encyclopedia. Today, a simple Internet search will return hundreds if not thousands of pages about even the most obscure topic. With the advent of Web 2.0, the amount of information stored in organized form exploded. At the time of this writing, the English version of the online encyclopedia Wikipedia includes more than 4.6 million articles containing more than a billion words. That is more than 30 times the number of words in the largest English-language encyclopedia ever published, the *Encyclopaedia Britannica*. The growth of the number of articles in Wikipedia (plotted in figure 3.9) has followed an accelerating curve, although it shows a tendency to decelerate as Wikipedia begins to cover a significant fraction of the world knowledge relevant to a large set of persons.

Wikipedia is just one of the many examples of services in which a multitude of users adds value to an ever-growing community, thus leading to a quadratic growth of utility. Other well-known examples are YouTube (which makes available videos uploaded by users), Flickr and Instagram (photos),

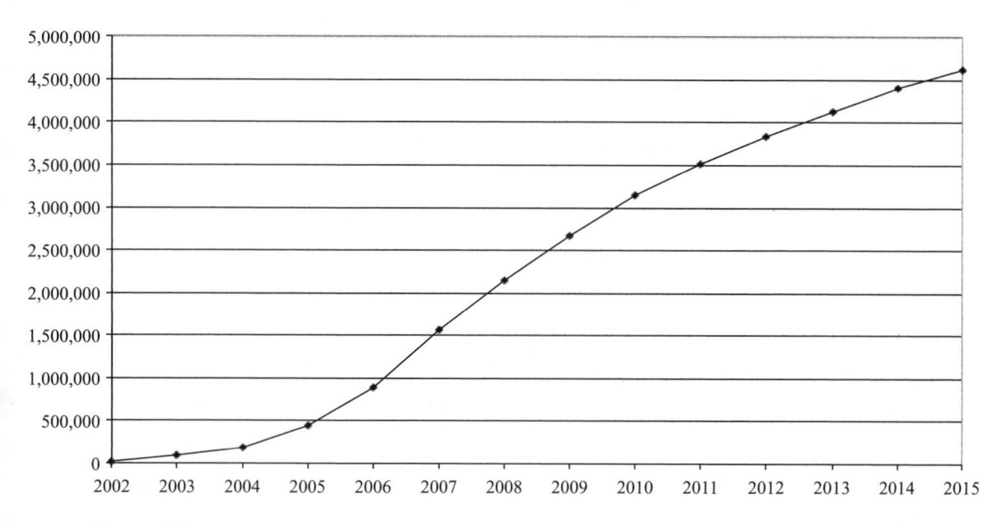

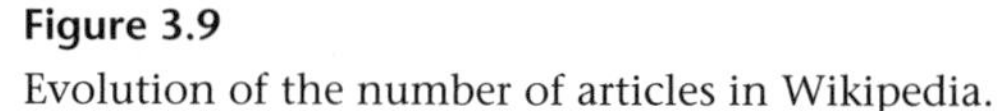

Figure 3.9
Evolution of the number of articles in Wikipedia.

and an array of social networking sites, of which the most pervasive is Facebook. A large array of specialized sites cater to almost any taste or persuasion, from professional networking sites such ass LinkedIn to Twitter (which lets users post very small messages any time, from anywhere, using a computer or a cell phone).

Electronic commerce—that is, use of the Web to market, sell, and ship physical goods—is changing the world's economy in ways that were unpredictable just a decade ago. It is now common to order books, music, food, and many other goods online and have them delivered to one's residence. In 2015 Amazon became the world's most valuable retailer, outpacing the biggest brick-and-mortar stores. Still, Amazon maintains physical facilities to store and ship the goods it sells. In a more radical change, Uber and Airbnb began to offer services (respectively transportation and lodging) using an entirely virtual infrastructure; they now pose serious threats to the companies that used to provide those services in the form of physical facilities (cabs and hotels).

Online games offer another example of the profound effects computers and the Internet can have on the way people live their daily lives. Some massively multiplayer online role-playing games (MMORPGs) have amassed very large numbers of subscribers, who interact in a virtual game world. The players, interconnected through the Internet, develop their game-playing activities over long periods of time, building long-term relationships and interacting in accordance with the rules of the virtual world. At the time of this writing, the most popular games have millions of subscribers, the size of the population of a medium-size country. Goods existing only in the virtual world of online games are commonly traded, sometimes at high prices, in the real world.

Another form of interaction that may be a harbinger of things to come involves virtual worlds in which people can virtually live, work, buy and sell properties, and pursue other activities in a way that mimics the real world as closely as technology permits. The best-known virtual-world simulator of this type may be Second Life, launched in 2003. Second Life has a parallel economy, with a virtual currency that can be exchanged in the same ways as conventional currency. Second Life citizens can develop a number of activities that parallel those in the real world. The terms of service ensure that users retain copyright for content they create, and the system provides simple facilities for managing digital rights. At present the user interface is still somewhat limited in its realism, since keyboard-based interfaces and relatively low-resolution computer-generated images are used to interact with the virtual world. Despite its relatively slow growth, Second Life now boasts about a million regular users.

This is a very brief and necessarily extremely incomplete overview of the impact of Internet technology on daily life. Much more information about these subjects is available in the World Wide Web, for instance, in Wikipedia. However, even this cursory description is enough to make it clear that there are millions of users of online services that, only a few years ago, simply didn't exist.

One defining aspect of present-day society is its extreme dependency on information and communication technologies. About sixty years ago, IBM was shipping its first electronic computer, the 701. At that time, only an irrelevant fraction of the economy was dependent on digital technologies. Telephone networks were important for the economy, but they were based on analog technologies. Only a vanishingly small fraction of economic output was dependent on digital computers.

Today, digital technologies are such an integral part of the economy that it is very difficult, if not impossible, to compute their contribution to economic output. True, it is possible to compute the total value created by makers of computer equipment, by creators of software, and, to a lesser extent, by producers of digital goods. However, digital technologies are so integrated in each and every activity of such a large fraction of the population that it isn't possible to compute the indirect contribution of these technologies to the overall economy. A number of studies have addressed this question but have failed to assign concrete values to the contributions of digital technologies to economic output.

It is clear, however, that digital technologies represent an ever-increasing fraction of the economy. This fraction rose steadily from zero about sixty years ago to a significant fraction of the economic output today. In the United States, the direct contribution of digital technologies to the gross domestic product (GDP) is more than a trillion dollars (more than 7 percent of GDP), and this fraction has increased at a 4 percent rate in the past two decades (Schreyer 2000)—a growth unmatched by any other industry in history. This, however, doesn't consider all the effects of digital technologies on everyday life that, if computed, would lead to a much higher fraction of GDP.

There is no reason to believe that the growth in the importance of digital technologies in the economy will come to a stop, or even that the rate of growth will reduce to a more reasonable value. On the contrary, there is ample evidence that these technologies will account for an even greater percentage of economic output in coming decades. It may seem that, at some point, the fraction of GDP due to digital technologies will stop growing. After all, some significant needs (e.g., those for food, housing, transportation, clothing, and energy) cannot be satisfied by digital technologies, and these needs will certainly account for some fixed minimum fraction of overall economic activity. For instance, one may assume, conservatively, that some fixed percentage (say, 50 percent) of overall economic output must be dedicated to satisfying actual physical needs, since, after all, there is only so much we can do with computers, cell phones and other digital devices. That, however, is an illusion based on the idea that overall economic output will, at some point, stagnate—something that has never happened and that isn't likely to happen any time soon. Although basic needs will have to be satisfied (at least for quite a long time), the potential of new services and products based on digital technology is, essentially, unbounded. Since the contribution of digital technologies to economic growth is larger than the contribution of other technologies and products, one may expect that, at some point in the future, purely digital goods will represent the larger part of economic output. In reality, there is no obvious upper limit on the overall contribution of the digital economy. Unlike physical goods, digital goods are not limited by the availability of physical resources, such as raw materials, land, or water. The rapid development of computer technology made it possible to deploy new products and services without requiring additional resources, other than the computing platforms that already exist. Even additional energy requirements are likely to be marginal or even non-existent as computers become more and more energy efficient.

This is nothing new in historical terms. Only a few hundred years ago, almost all of a family's income was used to satisfy basic needs, such as those for food and housing. With the technological revolutions, a fraction of this income was channeled to less basic but still quite essential things, such as transportation and clothing. The continued change toward goods and services that we deem less essential is simply the continuation of a trend that was begun long ago with the invention of agriculture.

One may think that, at some point, the fraction of income channeled into digital goods and services will cease to increase simply because people will have no more time or more resources to dedicate to the use of these technologies. After all, how many hours a day can one dedicate to watching digital TV, to browsing the Web, or to phone messaging? Certainly no more than 24, and in most cases much less. Some fraction of the day must be, after all, dedicated to eating and sleeping. However, this ignores the fact that the digital economy may create value without direct intervention of human beings. As we will see in chapter 11, digital intelligent agents may, on their behalf or on behalf of corporations, create digital goods and services that will be consumed by the rest of the world, including other digital entities. At some point, the fraction of the overall economic output actually attributable to direct human activity will be significantly less than 100 percent. To a large extent, this is already the case today. Digital services that already support a large part of our economy are, in fact, performed by computers without any significant human assistance. However, today these services are performed on behalf of some company that is, ultimately, controlled by human owners or shareholders. Standard computation of economic contributions ultimately attributes the valued added by a company to the company's owners.

In this sense, all the economic output generated today is attributable to human activities. It is true that in many cases ownership is difficult to trace, because companies are owned by other companies. However, in the end, some person or group of persons will be the owner of a company and, therefore, the generator of the economic output that is, in reality, created by very autonomous and, in some cases, very intelligent systems. This situation will remain unchanged until the day when some computational agent is given personhood rights, comparable to those of humans or corporations, and can be considered the ultimate producer of the goods or services. At that time, and only at that time, we will have to change the way we view the world economy as a product of human activity.

However, before we get to that point, we have to understand better why computers have the potential to be so disruptive, and so totally different from any other technology developed in the past. The history of computers predates that of the transistor and parallels the history of the discovery of electricity. Whereas the construction of computers that actually worked had to await the existence of electronic devices, the theory of computation has its own parallel and independent history.

4 The Universal Machine

Computers are now so ubiquitous that it is hard to imagine a time when the very concept of computer didn't exist. The word *computer*, referring to a person who carried out calculations, or computations, was probably used for the first time in the seventeenth century, and continued to be used in that sense until the middle of the twentieth century. A computer is someone or something that executes sequences of simple calculations—sequences that can be programmed. Several mechanical calculating devices were built in the sixteenth and seventeenth centuries, but none of them was programmable. Charles Babbage was the first to design and partially build a fully programmable mechanical computer; he called it the Analytical Engine.

The Analytical Engine

The Analytical Engine was a successor to the Difference Engine, a calculator designed and built by Babbage to automate the computation of astronomical and mathematical tables. Its workings were based on a tabular arrangement of numbers arranged in columns. The machine stored one decimal number in each column and could add the value of one cell in column $n + 1$ to that of one cell in column n to produce the value of the next row in column n. This type of repetitive computation could be used, for example, to compute the values of polynomials.

To understand the operation of the Difference Engine, it is useful to look at a concrete example. Consider, for instance, the function defined by the polynomial $P(x) = x^2 + 2x + 2$. Tabulating its values for the first few values of x, and the successive differences, yields the table shown here as table 4.1. As is clear from this table, the second-order differences for this polynomial are constant. This is a general characteristic of polynomials of degree 2. This

Table 4.1
Differences used by the Difference Engine.

x	$P(x)$	D1	D2
1	5		
2	10	5	
3	17	7	2
4	26	9	2
5	37	11	2
6	50	13	2

characteristic enables us to compute the value of the polynomial at any point without any multiplication. For instance, the value of the polynomial at $x = 7$ can be obtained by summing 2 (the constant in column D2) and the value 13 in column D1 and then adding the result and the value 50 in column $P(x)$ to obtain 65.

The Difference Engine was programmed by setting the initial values on the columns. Column $P(x)$ was set to the value of the polynomial at the start of the computation for a number of rows equal to the degree of the polynomial plus 1. The values in column D1 were obtained by computing the differences between the consecutives points of the polynomial. The initial values in the next columns were computed manually by subtracting the values in consecutive rows of the previous column. The engine then computed an arbitrary number of rows by adding, for each new row, the value in the last cell in a column to the value in the last cell in the previous column, starting with the last column, until it reached column $P(x)$, thereby computing the value of the polynomial for the next value of x.

The Difference Engine performed repetitive computations, but it couldn't be programmed to perform an arbitrary sequence of them; therefore, it cannot be considered a programmable computer. Babbage's second design, the Analytical Engine, was much more ambitious, and is usually considered the first truly programmable computer.

Babbage conceived carefully detailed plans for the construction of the Analytical Engine and implemented parts of it. (One part is shown here in figure 4.1.) The program and the data were to be given to the engine by means of punched cards (the same method used in the Jacquard loom, which was mentioned in chapter 2 as one of the results of the first industrial revolution). The Analytical Engine would generate its output using a

Figure 4.1
A part of the Analytical Engine now on exhibition at the Science Museum in London.

printer, but other output devices would also be available—among them a card puncher that could be used to generate future input.

The machine's memory was to hold 1,000 numbers of 50 decimal digits each. The central processing unit would perform the four arithmetic operations as well as comparisons. The fundamental operation of the Analytical Engine was to be addition. This operation and the other elementary operations—subtraction, multiplication, and division—were to be performed in a "mill". These operations were to be performed by a system of rotating cams acting upon or being actuated by bell cranks and similar devices.

The programming language to be employed by users was at the same level of abstraction as the assembly languages used by present-day

CPUs. An assembly language typically enables the programmer to specify a sequence of basic arithmetic and logic operations between different memory locations and the order in which these operations are to be performed. The programming language for the Analytical Engine included control instructions such as loops (by which the computer would be instructed to repeat specific actions) and conditional branching (by which the next instruction to be executed was made to depend on some already-computed value).

Had it been built, the Analytical Engine would have been a truly general-purpose computer. In 1842, Luigi Menabrea wrote a description of the engine in French; it was translated into English in the following year (Menabrea and Lovelace 1843), with annotations by Ada Lovelace (Lord Byron's daughter), who had become interested in the engine ten years earlier. In her notes—which are several times the length of Menabrea's description—Lovelace recognizes the machine's potential for the manipulation of symbols other than numbers:

> Again, it might act upon other things besides number, were objects found whose mutual fundamental relations could be expressed by those of the abstract science of operations, and which should be also susceptible of adaptations to the action of the operating notation and mechanism of the engine. Supposing, for instance, that the fundamental relations of pitched sounds in the science of harmony and of musical composition were susceptible of such expression and adaptations, the engine might compose elaborate and scientific pieces of music of any degree of complexity or extent.

She also proposes what can be considered the first accurate description of a computer program: a method for using the machine to calculate Bernoulli numbers. For these contributions, Ada Lovelace has been considered the first computer programmer and the first person to recognize the ability of computers to process general sequences of symbols rather than only numbers. The programming language Ada was named in her honor.

Technical and practical difficulties prevented Charles Babbage from building a working version of the Analytical Engine, although he managed to assemble a small part of it before his death in 1871. A working version of Babbage's Analytical Engine was finally built in 1992 and is on display in the Science Museum in London.

Turing Machines and Computers

After Babbage, the next fundamental advance in computing came from the scientist and mathematician Alan Turing, who proposed a model for an abstract symbol-manipulating device used to study the properties of algorithms, programs, and computations. (See Turing 1937.)

A Turing machine isn't intended to be a practical computing machine; it is an abstract model of a computing device. A deterministic Turing machine (which I will refer to in this chapter simply as a *Turing machine* or as a *machine*) consists of the following:

• An infinite tape divided into cells. Each cell contains a symbol from some finite alphabet. The alphabet contains a special blank symbol and one or more other symbols. Cells that have not yet been written are filled with the blank symbol.
• A tape head that can read and write symbols on the tape and move the tape left or right one cell at a time.
• A state register that stores the present state of the machine. Two special states may exist, an *accept* state and a *reject* state. The machine stops when it enters one of these states, thus defining whether the input has been accepted or rejected.
• A state-transition graph, which specifies the transition function. For each configuration of the machine, this graph describes what the tape head should do (erase a symbol or write a new one and move left or right) and the next state of the machine.

Figure 4.2 is a schematic diagram of a Turing machine. The state-transition graph describes the way the finite-state controller works. In this particular case, the state-transition graph has four states (represented by circles) and a number of transitions between them (represented by arrows connecting the circles). The controller works by changing from state to state according to the transitions specified. Each transition is specified by an input/output pair, the input and the output separated by a "slash." In the simple case illustrated in figure 4.2, the Turing machine has no input. Therefore, the transitions do not depend on the input, and only the outputs are shown. Figure 4.4 illustrates a slightly more complex case in which each transition is marked with an input/output pair.

Figure 4.2 is based on the first example proposed by Turing in his 1937 paper. The machine has four states: the initial state b (emphasized by a bold outline) and the states c, e, and f. Since the machine has no accept state and no reject state, it runs forever and simply writes the infinite sequence '0', ' ', '1', ' ', '0', ' ', '1' ... on the tape. Its behavior is simple to understand. The machine starts in state b, with the tape filled with blank symbols. The state-transition graph tells the machine to write a zero (P0), move to the right (R), and change to state c. In state c, the machine is told to move to the right (R) and not write anything, leaving a blank on the tape. The succeeding moves will proceed forever in essentially the same way. In figure 4.2, the controller is in state f and the machine will proceed to state b, moving the head to the right. Although this example is a very simple one, and the machine performs a very simple task, it isn't difficult to understand that much more complex tasks can be carried out. For example, a Turing machine can be programmed to add two binary numbers, written on the tape, and to write the result back on the tape. In fact, a Turing machine can be programmed

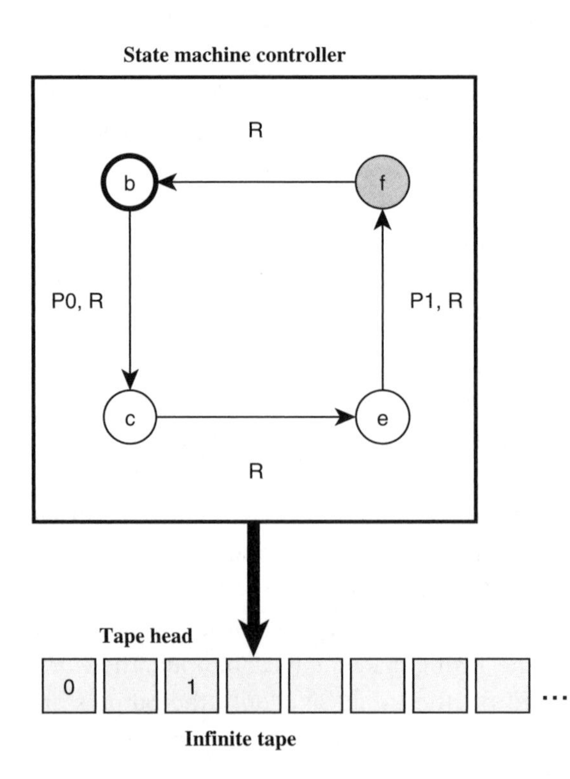

Figure 4.2
A Turing machine that writes an infinite succession of zeroes and ones with spaces in between.

to compute any number or to perform any symbol-manipulation task one can describe as a sequence of steps (an algorithm). It can compute the value of a polynomial at a certain point, determine if one equation has integer solutions, or determine if one image (described in the tape) contains another image (also described in another part of the tape).

In particular, a sufficiently powerful Turing machine can simulate another Turing machine. This particular form of simulation is called *emulation*, and this expression will be used every time one system simulates the behavior of another complete system in such a way that the output behavior is indistinguishable.

In the case of emulation of a Turing machine by another Turing machine, is it necessary to specify what is meant by "sufficiently powerful." It should be obvious that some machines are not powerful enough to perform complex tasks, such as simulating another machine. For instance, the machine used as an example above can never do anything other than write zeroes and ones. Some Turing machines, however, are complex enough to simulate any other Turing machine. A machine A is universal if it can simulate any machine B when given an appropriate description of machine B, denoted <B>, written on the tape together with the contents of the tape that is the input to machine B. The notation <B> is used to designate the string that is to be written on the tape of machine A in order to describe machine B in such a way that machine A can simulate it.

The idea of universal Turing machines gives us a new way to think about computational power. For instance, if we want to find out what can be computed by a machine, we need not imagine all possible machines. We need only imagine a universal Turing machine and feed it all possible descriptions of machines (and, perhaps, their inputs). This reasoning will be useful later when we want to count and enumerate Turing machines in order to assess their computational power.

A Turing machine is an abstraction, a mathematical model of a computer, which is used mainly to formulate and solve problems in theoretical computer science. It is not meant to represent a realistic, implementable, design for an actual computer. More realistic and feasible computer models, that are also implementable designs, have become necessary with the advent of actual electronic computers. The most fundamental idea, which resulted in large part from the work of Turing, was the stored-program computer, in which a memory, instead of a tape, is used to store the sequence of instructions that represents a specific program. In a paper presented in 1946, in the United Kingdom, to the National Physical Laboratory Executive Committee, Turing presented the first reasonably finished design of a stored-program computer, a device he called the Automatic Computing Engine (ACE). However, the EDVAC design of John von Neumann (who was familiar with Turing's theoretical work) became better known as the first realistic proposal of a stored-program computer.

John von Neumann was a child prodigy, born in Hungary, who at the age of 6 could divide two eight-digit numbers in his head. In 1930 he received an appointment as a visiting professor at Princeton University. At Princeton he had a brilliant career in mathematics, physics, economics, and computing. Together with the theoretical physicists Edward Teller and Stanislaw Ulam, he developed some of the concepts that eventually led to the Manhattan Project and the atomic bomb.

One of von Neumann's many contributions to science was what eventually became known as the von Neumann architecture. In "First Draft of a Report on the EDVAC" (von Neumann 1945), he proposes an architecture for digital computers that includes a central processing unit, a control unit, a central memory, external mass storage, and input/output mechanisms. The central processing unit (CPU) contains a control unit (CU) and an arithmetic and logic unit (ALU). The ALU computes arithmetical and logical operations between the values stored in the registers in the CPU. The control unit includes a program counter, which points to the specific point in memory that contains the current instruction and is used to load the instruction register (one of the registers in the CPU) with the code of the operation to be performed. The memory, which is used to store both data and instructions, is accessed using the registers in the CPU. Finally, external mass storage is used to store permanently instructions and data, for future use, while input and output devices enable the computer to communicate with the external world. The memory and the external mass storage replace the tape in the Turing machine, while the internal structure of the central processing unit makes it more efficient in the manipulation of information stored in memory and in permanent storage.

In the so-called von Neumann architecture, the same pathway is used both to load the program instructions and to move the working data between the CPU and the memory—a restriction that may slow down the operation of the computer. The so-called Harvard architecture, named after

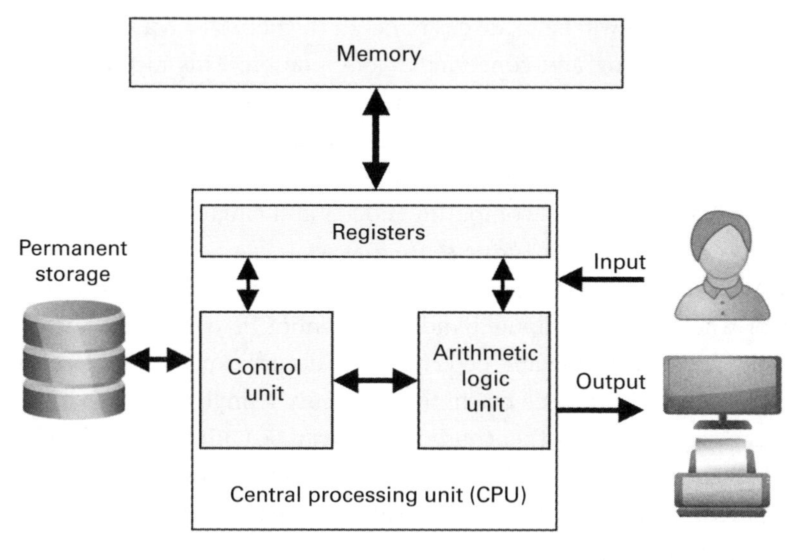

Figure 4.3
Von Neumann architecture of a stored-program computer.

the Harvard Mark I relay-based computer, uses separate pathways (or buses) to move the working data between the CPU and the memory and to load the program instructions. Some modern computers, including computers dedicated to digital signal processing, use the Harvard architecture; others use a combination of the Harvard architecture and the von Neumann architecture.

Modern machines have multiple buses and multiples CPUs, and use complex mechanisms to ensure that different CPUs can access the available memory, which may be organized in many different ways. However, all modern computers work in essentially the same way and are, essentially, very fast implementations of the stored-program computers designed by Alan Turing and John von Neumann.

Computability and the Paradoxes of Infinity

The advantage of a Turing machine over more complex computer models, such as the stored-program computer, is that it makes mathematical analysis of its properties more straightforward. Although a Turing machine may seem very abstract and impractical, Turing proved that, under some assumptions (which will be made clear later in the book) such a machine is capable of performing any conceivable computation. This means that a Turing machine is as powerful as any other computing device that will ever be built. In other words, a Turing machine is a universal computer, and it can compute anything that can be computed.

We now know almost all computing models and languages in existence today are equivalent in the Turing sense that what can be programmed in one of them can be programmed in another. This leads to a very clear definition of what can be computed and what cannot be computed. Such a result may seem surprising and deserves some deeper explanation. How is it possible that such a simple machine, using only a single tape as storage, can be as powerful as any other computer that can be built? Exactly what is meant by "anything that can be computed"? Are there things that cannot be computed? To dwell further on the matter of what can and what cannot be computed, we must answer these questions clearly and unequivocally. To do this, we use the concept of *language*.

A language is defined as a set of sequences of symbols from some given alphabet. For instance, the English language is, according to this definition, the set of all possible sequences of alphabet symbols that satisfy some specific syntactic and semantic rules. For another example, consider the language L_1 = {0, 00, 000, 0000, 00000, ... }, which is constituted by all strings consisting only of zeros. There are, of course, an infinite number of such strings and, therefore, this language has infinitely many strings.

A language L is said to be *decidable* if there exists a Turing machine that, given a string written on the tape, stops in the accept state if the string is in L and stops in the reject state if the string is not in L. Language L_1 is decidable, since there is a machine (illustrated here in figure 4.4, performing two different computations) that does exactly that. The two parts of figure 4.4 illustrate two computations performed by this Turing machine, showing the state of the machine after two inputs—0001 (figure 4.4a) and 000 (figure 4.4b)—have been processed. In figure 4.4a, the machine has just rejected the string, with the controller having moved into the reject state after seeing a 1 in the fourth cell of the tape. In figure 4.4b, the machine has just accepted the string, after seeing a space on the tape following a string of zeroes.

A language L is said to be *recognizable* when there exists a machine that stops in the accept state when processing a string in L but may fail to stop

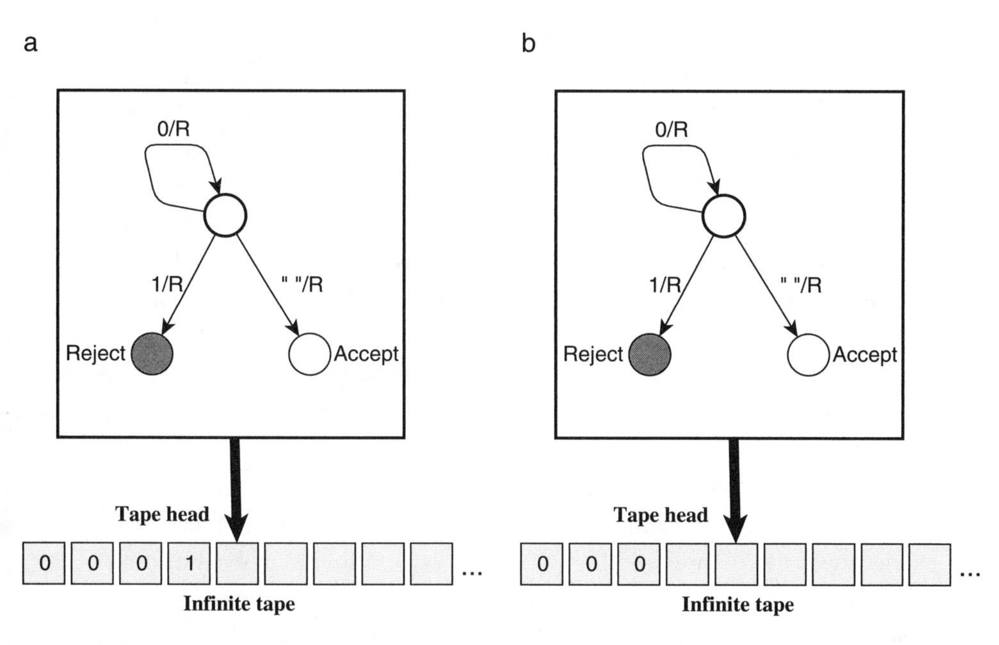

Figure 4.4
A Turing machine that accepts the language L_1.

in the reject state if the string is not in L. Decidable languages are, therefore, "simpler" than recognizable languages, since a Turing machine can definitely answer the question "Does a string belong to L?" in a decidable language, but can only answer in one direction—if the answer is yes—if L is a recognizable language.

What is the relationship between languages and problems? Informally, we use the word *problem* to refer to a question that has a specific answer. If one restricts the analysis to problems that can be answered either Yes or No, it is possible to define a correspondence between problems and languages. Each instance of the problem is encoded as a string, and all the strings that correspond to instances with Yes answers define the language that corresponds to the problem. The other strings either fail to correspond to any instance or correspond to instances with No answers. A problem that corresponds to a decidable language is said to be decidable. Otherwise, a problem is said to be undecidable or, equivalently, non-computable.

In more practical and useful terms, undecidable problems are those for which there is no algorithm that is guaranteed to always find a solution. This means that no fixed set of instructions, executing in finite time by a computer, is guaranteed to lead to a solution. The fact that some problems are undecidable was one of Alan Turing's major contributions to the theory of computation. Alonzo Church (1936) used a different formalism and arrived at a different definition of computability, now known to be equivalent to Turing's definition. Kurt Gödel's (1931) famous incompleteness theorem dealing with the existence of mathematical truths that cannot be demonstrated or derived from a fixed set of axioms is also ultimately equivalent to Turing's result that there are problems that are undecidable, or non-computable.

It is important to understand why undecidable problems necessarily exist. To do so, we have to go a little deeper into the concepts of computability and infinity. To prove that undecidable problems necessarily exist, we must first become more familiar with the concept of infinity and with the fact that not all infinite sets have the same number of elements. Consider the set of natural numbers, {1, 2, 3, ... }, and the set of even numbers, {2, 4, 6, ... }, both of infinite cardinality. Do they have the same cardinality, the same number of elements? One may be tempted to say there are more natural numbers than even numbers, because every even number is also a natural number, and the natural numbers also include all the odd numbers. However, as table 4.2 shows, we can create a one-to-one correspondence between the set of natural numbers and the set of even numbers.

Now, it is eminently reasonable to argue that if a one-to-one correspondence can be established between the elements of two sets, then the two sets will have the same number of elements—that is, the same cardinality. This is exactly the mathematical definition of cardinality that was proposed in 1874 by the mathematician Georg Cantor, and the only one that can reasonably be used when the number of elements in the sets is infinite.

Table 4.2
Lists of natural and even numbers, in one-to-one correspondence.

Natural numbers	Even numbers
1	2
2	4
3	6
4	8
5	10
...	...

There are, therefore, as many even numbers as natural numbers. Consider now the set of all positive rational numbers, the numbers that can be written as the ratio of two positive integers. These numbers include 0.5 (1/2), 0.25 (1/4), 0.3333... (1/3), and 1.428571428571... (10/7), among infinitely many others. Certainly there are more positive rational numbers than natural numbers, since, intuition would tell us, an infinite number of rational numbers exists between any two integers. In fact, the situation is even more extreme, because an infinite number of rational numbers exists between any two chosen rational numbers.

Alas, intuition fails us again. The cardinality of the set of rational numbers is no larger than the cardinality of the set on integers. Using a somewhat smarter construction, it is possible to build a list of all rational numbers. If an ordered list of all rational numbers can be built, then they can be put in one-to-one correspondence with the natural numbers, and the cardinality of the set of rational numbers is the same as the cardinality of the set of naturals.

To list the positive rational numbers, let us first consider those for which the numerator plus the denominator sum to 2, then those for which they sum to 3, then those for which they sum to 4, and so on. We should be careful to not include the same number twice. For instance, since 1/2 is already there, we should not include 2/4. Listing all positive rational numbers using this approach will lead to table 4.3. It should be clear that every positive rational number will show up somewhere in this list and that there is a one-to-one correspondence between the rational numbers and the natural numbers. This demonstrates they have the same cardinality.

Table 4.3
Lists of natural and rational numbers, in one-to-one correspondence.

Natural numbers	Rational numbers
1	1/1
2	1/2
3	2/1
4	1/3
5	3/1
6	1/4
7	2/3
...	...

Like the list of even numbers, the set of positive rational numbers is said to be *countable*, since it can be put in correspondence with the natural numbers (the set we use to count). It would seem that, by using similar tricks, we can create tables like those above for just about any infinite set. As long as the elements of the set can be listed in some specific order, the set will be no larger than the set of natural numbers. This would mean that all infinite sets have the same cardinality, called $\aleph_0$ (aleph zero)—the cardinality of the set of natural numbers. Alas, this is not the case, because, as Cantor demonstrated, there are sets whose elements cannot be put in an ordered list. These sets have higher cardinality than the set of natural numbers, as can be demonstrated with a simple argument.

Imagine that you could build a list of the real numbers between 0 and 1. It would look something like table 4.4, where each α_i, each β_i, each γ_i, and so on is a specific digit. Now consider the number $0.\alpha\beta\gamma\delta\varepsilon$ ..., where $\alpha \neq \alpha_1$, $\beta \neq \beta_2$, $\gamma \neq \gamma_3$, and so on. For instance, this number has digit α equal to 5 unless $\alpha_1 = 5$, in which case it has digital α equal to 6 (or any other digit different from 5). In the same way, digit β equals 5 unless $\beta_2 = 5$, and in that case β equals 6. Proceed in the same way with the other digits. Such a number, $0.\alpha\beta\gamma\delta\varepsilon$..., necessarily exists, and it should be clear it is nowhere in the list. Therefore, not all real numbers between 0 and 1 are in this list. The only possible conclusion from this reasoning is that a complete list of real numbers, even only those between 0 and 1, cannot be built. The cardinality of the set of these numbers is, therefore, larger than the cardinality of the set of natural numbers.

Now consider some fixed alphabet Σ, and consider the set of all possible finite strings that can be derived from this alphabet by concatenating symbols in Σ. We will call this language Σ^*. This set is countable, since we can list all the strings in the same way we can list all rational numbers. For

Table 4.4
Hypothetical lists of natural and real numbers, in one-to-one correspondence.

Natural numbers	Real numbers
1	$0.\alpha_1\ \beta_1\ \gamma_1\ \delta_1\ \varepsilon_1$ ...
2	$0.\ \alpha_2\ \beta_2\ \gamma_2\ \delta_2\ \varepsilon_2$ ...
3	$0.\ \alpha_3\ \beta_3\ \gamma_3\ \delta_3\ \varepsilon_3$ ...
4	$0.\ \alpha_4\ \beta_4\ \gamma_4\ \delta_4\ \varepsilon_4$ ...
...	....

instance, we can build such a list by ordering the set, starting with the strings of length 0 (there is only one—the empty string, usually designated by ε), then listing the strings of length 1, length 2, and so on. Such a list would look something like (s_1, s_2, s_3, s_4, ...). Any language L defined over the alphabet Σ is a subset of Σ*, and can be identified by giving an (infinite) string of zeroes and ones, which identifies which strings in the list belong to L. For instance, if Σ were equal to {0,1} the list of strings in Σ* would look like (ε, 0, 1, 00, 01, 10, 11, 000, 001, ...). As an example, consider the languages L_2, consisting of the strings starting with a 0, and the language L_3, consisting of the strings having exactly an even number of zeroes. Each one of these languages (and any language defined over Σ*) can be uniquely defined by an (infinite) string of zeroes and ones, a string that has a 1 when the corresponding string in the list of strings in Σ* belongs to the language and has a 0 when the corresponding string doesn't belong to the language. Figure 4.5 illustrates this correspondence.

At the top of figure 4.5 we have all strings that can be written with the alphabet Σ, written in some specific order. The order doesn't matter as long as it is fixed. At the bottom of the L_2 box, we have all the strings that belong to L_2. In our example, language L_2 includes all the strings that begin with a 0. At the top of the L_2 box, we have the infinite string of zeroes and ones that characterizes L_2. In the L_3 box, we have the same scheme. Recall that language L_3 includes all strings with exactly an even number (including none) of zeroes. This set of strings is shown at the bottom of the box; the infinite string of zeroes and ones that characterizes L_3 is shown at the top of the box. It should be clear that there is a one-to-one correspondence between each language defined over some alphabet Σ and each infinite

Σ*	(ε,	0,	1,	00,	01,	10,	11,	000,	001,	...)
	0	1	0	1	1	0	0	1	1	...
L_2	(,	0,	,	00,	01,	,	,	000,	001,	...)
	1	0	0	1	0	0	0	0	1	...
L_3	(ε,	,	,	00,	,	,	,	,	001,	...)

Figure 4.5
An illustration of the equivalence between infinite strings of zeroes and ones and languages defined over an alphabet.

string of zeroes and ones, defined as above. Now, as it turns out, the set of all infinite strings of zeroes and ones is not countable. We can never build a list of the elements in this set, for the same reasons we could not build a set of the real numbers. This impossibility can be verified by applying the same diagonalization argument that was used to demonstrate that the set of reals is not countable. This implies that the set of all languages over some alphabet Σ is not countable.

It turns out the set of all Turing machines that work on the alphabet Σ is countable. We can build the (infinite) list of all Turing machines listing first those with zero states, then the ones with one state, then those with two states, and so on. Alternatively, we may think that we have a universal Turing machine that can simulate any other machine, given its description of the tape. Such a universal Turing machine can emulate any Turing machine, and the description of any of these machines is a fixed set of symbols. These descriptions can be enumerated. Therefore, we can build a list of all machines. It will look something like {TM_1, TM_2, TM_3, TM4, … }.

It is now clear that the set of all languages over some alphabet cannot be put in one-to-one correspondence with the set of Turing machines. Since a Turing machine recognizes one language at most (and some Turing machines do not recognize any language), we conclude that some languages are not recognized by any machine, since there are more languages than Turing machines. In equivalent but less abstract terms, we could say that the set of all programs is countable and the set of all problems is not countable. Therefore, there are more problems than programs. This implies there are problems that cannot be solved by any program.

This reasoning demonstrates there are uncountably many languages that cannot be recognized by any Turing machine. However, it doesn't provide us with a concrete example of one such language. To make things a bit less abstract, I will describe one particular undecidable problem and provide a few more examples that, despite their apparent simplicity, are also undecidable.

Consider the following problem, called the halting problem: Given a machine M and a string w, does M accept w? We will see that no Turing machine can decide this problem. Suppose there is a decider for this problem (call it H) that takes as input a description of M, <M>, and the string w, <M>:w. <M>:w represents simply the concatenation of the description of M, <M>, and the string w. Machine H accepts the input (i.e., halts in the accept state) if and only if M accepts w. If M doesn't accept w, then machine

H halts in the reject state. Clearly, H is a machine that can decide whether or not another machine, M, accepts or not a given input, w.

If such a decider, H, exists, we can change it slightly to build another Turing machine, Z, that is based on H (with a few modifications) and works in the following way: On input <X>, a description of machine X, Z runs like H on input <X>:<X>, the concatenation of two descriptions of machine X. Then, if H accepts, Z rejects, and if H rejects, Z accepts. Z can easily be built from H by switching the accept and reject labels on the halting states of H. Note that running H on input <X>:<X> is nothing strange. H is expecting as input a description of a machine and a string w. Z, a trivial modification of H, runs on the description of X, <X>, and a string that, in this case, corresponds exactly to the description of X, <X>. In other words, Z is a machine that can decide whether a machine X accepts its own description as input. If X accepts <X>, then Z stops in the reject state. Otherwise, Z stops in the accept state.

We can, obviously, apply Z to any input. What happens when we run Z on input <Z>? Z must accept if Z rejects <Z> and must reject if Z accepts <Z>. This is a contradiction because no such machine can exist. The unique way out of the contradiction is to recognize that Z cannot exist. Since Z was built in a straightforward way from H, the conclusion is that H, a decider for the halting problem, cannot exist. Because no decider for this problem exists, the halting problem is undecidable. In less formal but perhaps more useful terms, you cannot build a program A that, when given another program B and an input to B, determines whether B stops or not.

Since the halting problem is, itself, rather abstract, let us consider another slightly more concrete example, which comes from the mathematician David Hilbert. At the 1900 conference of the International Congress of Mathematicians in Paris, Hilbert formulated a list of 23 open problems in mathematics (Hilbert 1902). The tenth problem was as follows: Given a Diophantine equation specified by a polynomial with more than one variable and integer coefficients, is there a set of integer values of the variables that make the polynomial evaluate to 0? For example, given the polynomial

$$xy^3 + x^5 - 4x^2y^3 + 7,$$

is it possible to select integer values for x and y that make it evaluate to 0? Hilbert asked for a finite sequence of steps (i.e., an algorithm) that would be guaranteed to succeed in determining whether a solution exists for an equation of this type, always answering either Yes or No. Without loss of generality, and to simplify the discussion, we may assume the integer values that will be attributed to x and y are positive integers.

In terms of Turing machines, this problem has to be represented by a language, which is, as we know, a set of strings. Each string in the language represents an encoding of a particular instance of the problem—in this case, a Diophantine equation. The language that corresponds to Hilbert's tenth problem will therefore be a set of strings, each of which encodes a particular instance of the problem. For instance, we may decide that the string $\{x, 2, y, 1, 0, +, 1, 5\}$ encodes the polynomial $x^2y^{10} + 15$. This language will have infinitely many strings, since the number of polynomials is infinite.

Given what we know about Turing machines, we may be able to think of a way to solve this. We can start with the input polynomial written on the tape. The machine then tries all possible combinations for the values of x and y in some specific sequence—for example, (0,0), (0,1), (1,0), (0,2), (1,1), (2,0), ... For each combination of values, the machine evaluates the polynomial. If the value is 0, it halts. Otherwise, it proceeds to the next combination. Such an approach leads to a Turing machine that halts if a solution is found, but that will run forever if there is no solution to the problem. One may think that it should be relatively simple to change something in order to resolve this minor difficulty. However, such is not the case. In fact, we now know, through the work of Martin Davis, Yuri Matiyasevich, Hilary Putnam, and Julia Robinson, that this problem is undecidable (Davis, Putnam, and Robinson 1961; Davis, Matiyasevich, and Robinson 1976). This means that no Turing machine can ever be built to recognize the strings in this language—that is, a machine that would always stop, accepting when the string corresponds to a Diophantine equation with an integer solution and rejecting when the string corresponds to an equation without integer solutions. It must be made clear that, for some specific equations, it may be possible to determine whether or not they have solutions. The general problem, however, is undecidable, since for some equations it isn't possible to determine whether they do or don't have integer solutions.

One might think that only very abstract problems in logic or mathematics are undecidable. Such is not the case. I will conclude this section with an example of an apparently simpler problem that is also undecidable: the Post Correspondence Problem (PCP).

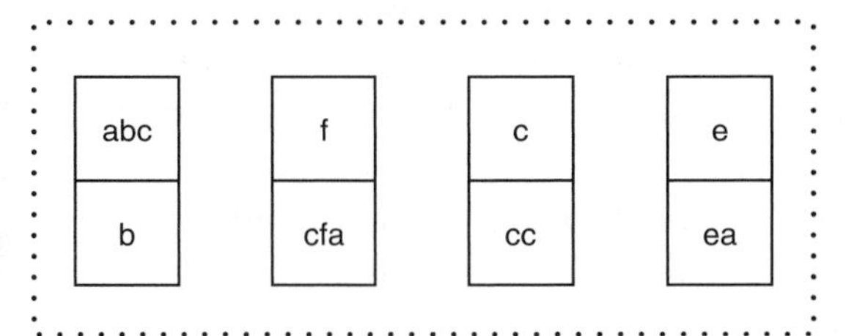

Solution

e	abc	f	abc	c
ea	b	cfa	b	cc

Figure 4.6
Illustration of the Post Correspondence Problem, known to be undecidable.

Suppose you are given a set of dominos to use as building blocks. Each half of a domino has one particular string inscribed. For instance, one domino may have abc at the top and b at the bottom. Given a set of these dominoes, and using as many copies of each domino as necessary, the Post Correspondence Problem consists in determining if there is a sequence of dominoes that, when put side by side, spells the same string at the top and at the bottom. For instance suppose you are given the set {abc/b, f/cfa, c/cc, e/ea}. Is there a sequence of (possibly repeating) dominoes that spells the same string at the top and at the bottom? The answer, in this case, is Yes. The sequence {e/ea, abc/b, f/cfa, abc/b, c/cc} spells the string "eabcfabcc" at the top and at the bottom, as shown in figure 4.6.

It turns out that the problem that corresponds to the apparently simple puzzle just discussed is also undecidable (Post 1946), which implies that no algorithm is guaranteed to always work. If an algorithm to solve this problem exists, then it can also be used to solve the halting problem (Sipser 1997), which, by contradiction, implies that no such algorithm exists. Note that in some cases it may be possible to answer in a definite way, either positively or negatively, but it isn't possible to devise a general algorithm for this problem that is guaranteed to always stop with a correct Yes or No answer.

Algorithms and Complexity

We have learned, therefore, that some problems are undecidable—that is, no algorithm can be guaranteed to find a solution in all cases. The other problems, which include most of the problems we encounter in daily life, are decidable, and algorithms that solve them necessarily exist. In general, algorithms are designed in abstract terms as sequences of operations, without explicit mention of the underlying computational support (which can be a stored-program computer, a Turing machine, or some other computational model). The algorithms can be more or less efficient, and can take more or less time, but eventually they will stop, yielding a solution.

I have already discussed algorithms and their application in a number of domains, ranging from electronic circuit design to artificial intelligence. Algorithm design is a vast and complex field that has been at the root of many technological advances. Sophisticated algorithms for signal processing make the existence of mobile phones, satellite communications, and secure Internet communications possible. Algorithms for computer-aided design enable engineers to design computers, cars, ships, and airplanes. Optimization algorithms are involved in the logistics and distribution of almost everything we buy and use. In many, many, cases, we use algorithms unconsciously, either when we use our brain to plan the way to go to the nearest coffee shop or when we answer a telephone call.

An algorithm is a sequence of steps that, given an input, achieves some specific result. To make things less abstract, I will give a few concrete examples of commonly used algorithms in a specific domain: the domain of cities, maps, and road distances. From the point of view of these algorithms, the map of cities, roads, and distances will be represented by a graph. A graph is a mathematical abstraction used to represent many problems and domains. A graph consists of a set of nodes and a set of edges. The edges connect pairs of nodes. Both the nodes and the edges can have weight, and the edges can be either directed or non-directed, depending on the nature of the problem. The number of edges that connects to a given node is called the *degree* of the node. If the graph is directed, the number of incoming vertices is called the *in-degree* and the number of outgoing vertices is called the *out-degree*.

Let the graph shown in figure 4.7 represent the roads and the distances between two nearby cities in France. Now consider the familiar problem of finding the shortest distance between two cities on a map. Given any two cities, the algorithm should find the shortest path between them. If, for example, one wants to go from Paris to Lyon, the shortest way is to go through Dijon, for a total distance of 437 kilometers (263 from Paris to Dijon plus 174 from Dijon to Lyon). It is relatively easy to find the shortest path between any two cities, and there are a number of

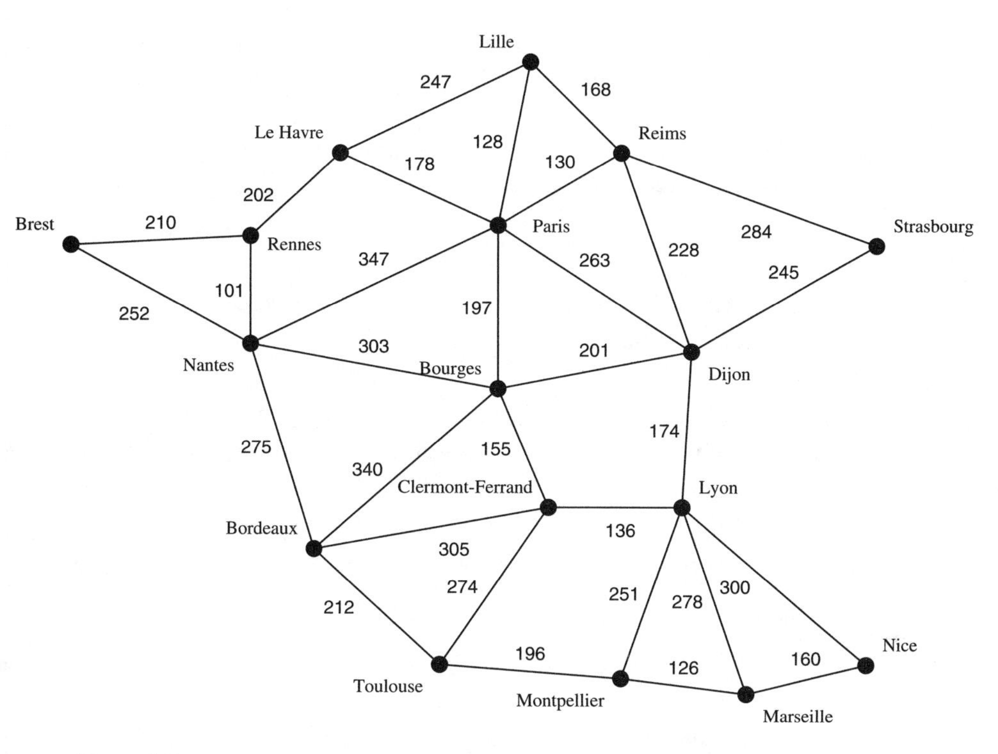

Figure 4.7
A graph of the distances between French cities.

efficient algorithms that can compute it efficiently. The first such algorithm, proposed by Edsger Dijkstra in 1956, finds the shortest path to any node from a single origin node. (See Dijkstra 1959.) As the graph gets larger, the problem becomes more challenging and the execution time of the algorithm increases; however, the increases are progressive, and very large graphs can be handled.

It is also easy to answer another question about this graph: What is the set of roads with shortest total length that will keep all the cities connected? This problem, known as the *minimum spanning tree,* has many practical applications and has even been used in the study of brain networks (discussed in chapter 9). An algorithm to solve this problem was first proposed as a method of constructing an efficient electricity network for Moravia (Borůvka 1926).

The minimum spanning tree can be found by a number of very efficient algorithms. Prim's algorithm is probably the simplest of them. It simply selects to include in the spanning tree, in order, the shortest road still unselected that doesn't create, by being chosen, a road loop (Prim 1957). The algorithm stops when all the cities are connected. If applied to the road map shown in figure 4.7, it yields the tree shown in figure 4.8.

Other, more challenging problems can be formulated on this graph. Imagine you are a salesman and you need to visit every city exactly once and then return to your starting city. This is called the *Hamiltonian cycle problem* after William Rowan Hamilton, who invented a game (now known as Hamilton's Puzzle) that involves finding a cycle in the edge graph of a dodecahedron—that is, a platonic regular solid with twelve faces. Finding a Hamiltonian cycle in a graph becomes complex rapidly as the size of the graph increases. If you want to visit all the cities as quickly as is possible, by the shortest possible route, the challenge is called the *traveling salesman problem*. Again, the problem is easy when there are only a few cities, but becomes harder very rapidly when there are many cities.

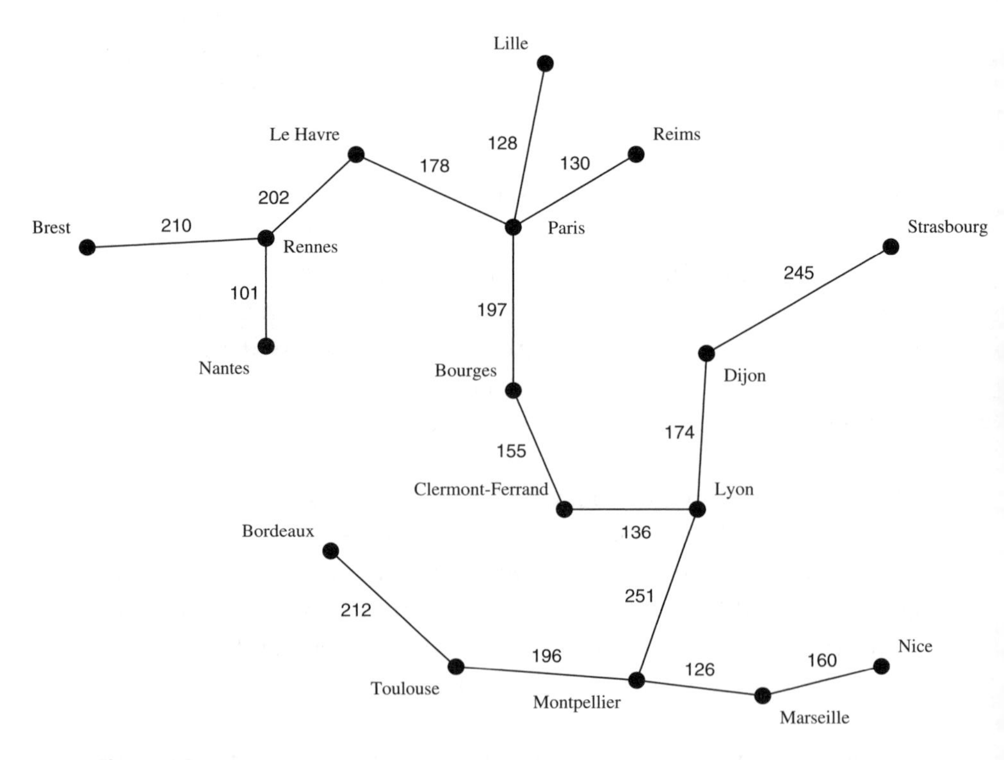

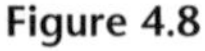
Figure 4.8
The minimum spanning tree for the graph in figure 4.7.

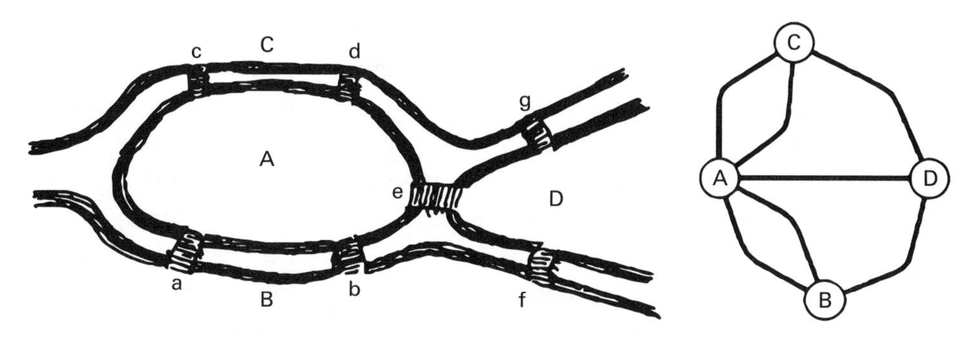

Figure 4.9
A drawing (by Leonhard Euler) and a graph representation of the Seven Bridges of Königsberg problem.

Another simple problem that can be formulated in graphs is the *Eulerian path problem*, the name of which recalls Leonhard Euler's solution of the famous Seven Bridges of Königsberg problem in 1736. The city of Königsberg had seven bridges linking the two sides of the Pregel River and two islands. (See figure 4.9.) The problem was to find a walk through the city that would cross each bridge once and only once. The problem can be represented by a graph with a node standing for each island and each margin of the river and with each edge representing a bridge. (In this case the graph is called a *multi-graph*, because there are multiple edges between the same two nodes.) A path through a graph that crosses each edge exactly once is called an Eulerian path. An Eulerian path that returns to the starting point is called an *Eulerian cycle*. By showing that a graph accepts an Eulerian path if and only if exactly zero or two vertices have an odd degree, Euler demonstrated it was not possible to find a walk through Königsberg that would cross each bridge exactly once. If there are zero vertices with an odd degree, then all Eulerian paths are also Eulerian cycles. (We will encounter Eulerian paths again in chapter 7.)

Designing algorithms is the art of designing the sequence of steps that, if followed, arrives at a correct solution in the most efficient way. The efficiency of algorithms is basically the time they take to be executed when run on a computer. Of course, the actual time it takes to run an algorithm depends on the specific computer used. However, even the fastest computer will take a very long time to execute an algorithm that is very inefficient, in a very large problem, and even a relatively slow computer can execute an efficient algorithm in a problem of reasonable size. This is because the run time grows very rapidly with the size of the problem when the computational complexity of the algorithm is high. The computational complexity of algorithms is measured by the way they scale with the dimension of the problem. Usually the dimension of the problem is the size of the input needed to describe the problem. In our examples, the size of the graph can be the number of nodes (locations) plus the number of edges (roads or bridges) in the problem description. Technically we should also worry about the length of the description of each node and edge; in many cases, however, we may assume that this factor is not critical, since each of them can be described by a small and constant number of bits. For instance, if the size of the problem is N and the time of execution of an algorithm grows linearly with the size of the problem, we will say that the complexity of the algorithm is on the order of N. On the other hand, the run time can also grow with the square of the size of the problem; in that case, the complexity of the algorithm is on the order of N^2, and the run time of the algorithm quadruples when the size of the problem doubles. However, the run time of an algorithm can grow even faster. For instance, when the complexity is on the order of 2^N, if the size of the problem increases by only one, the run time of the algorithm doubles.

There is a big difference between the growth rate of algorithms that have polynomial complexity (for example, N^2) and those that have exponential complexity (for example, 2^N). The reason is that algorithms with exponential complexity become too time consuming, even if one uses fast computers and deals only with small problems. On the other hand, algorithms with polynomial complexity can usually solve very large problems without spending inordinate amounts of time.

Imagine a computer that can execute a billion (10^9) operations per second and three different algorithms that take, respectively, N, N^2, and 2^N operations to solve a problem of size N. It is easy to see that, for a problem of size 1,000, an algorithm that requires N operations will spend one microsecond of computer time. Nonetheless, it is instructive to compute how the execution time grows with N for each of the three algorithms. Table 4.5 gives the time it would take for this hypothetical computer (comparable to existing computers) to solve this problem with algorithms of varying complexity. Entries in the table contain the symbol ∞ if the run time is larger than the known age of the universe (about 14 billion years). As

Table 4.5
Execution times for algorithms of varying complexity. Here ∞ stands for times that are longer than the age of the universe.

	Algorithm complexity		
Problem size	N	N^2	2^N
10	10 nanoseconds	100 nanoseconds	1 microsecond
20	20 nanoseconds	400 nanoseconds	1 millisecond
50	50 nanoseconds	2.5 microseconds	13 days
100	100 nanoseconds	10 microseconds	∞
200	200 nanoseconds	40 microseconds	∞
500	500 nanoseconds	250 microseconds	∞
1000	1 microsecond	1 millisecond	∞
1,000,000	1 millisecond	17 minutes	∞
1,000,000,000	1 second	32 years	∞

becomes clear from the table, even moderately large problems take too long to be solved by algorithms with exponential complexity. This is, again, a consequence of the now-familiar properties of exponential functions, which grow very rapidly even when starting from a very low base.

On the other hand, algorithms with polynomial complexity can handle very large problems, although they may still take a long time if the problem is very large and the degree of the polynomial is higher than one. For these reasons, polynomial time algorithms are usually called *efficient algorithms*, even though, in some cases, they may still take a long time to terminate. Problems are called tractable if algorithms that solve them in polynomial time are known, and are called intractable otherwise. This division into two classes is, in general, very clear, because there is a class of problems for which polynomial time algorithms exist, and there is another class of problems for which no polynomial time algorithms are known.

The problem of finding the shortest path between two cities and the traveling salesman problem belong to these two different classes of complexity. In fact, a number of polynomial time algorithms are known that solve the shortest-path problem (Dijkstra 1959; Floyd 1962) but only exponential time algorithms are known that solve the traveling salesman problem (Garey and Johnson 1979). The Eulerian path problem can also be solved in polynomial time, with an algorithm that grows only linearly with the size of the graph (Fleischner 1990).

One might think that, with some additional research effort, a polynomial time algorithm could be designed to solve the traveling salesman problem. This is, however, very unlikely to be true, because the traveling salesman problem belongs to a large class of problems that are, in a sense, of equivalent complexity. This class, known as NP, includes many problems for which no efficient solution is known to exist. The name NP stands for *Non-deterministic Polynomial-time,* because it is the class of decision problems a non-deterministic Turing machine can solve in polynomial time. A non-deterministic Turing machine is a machine that, at each instant in time, can take many actions and (non-deterministically) change state to many different states. A decision problem is a problem that is formulated as a question admitting only a Yes or a No answer.

Non-deterministic Turing machines are even stranger and less practical than the deterministic Turing machines described before. A non-deterministic Turing machine is only a conceptual device. A non-deterministic Turing machine cannot be built, because at each instant in time it could move to several states and write several symbols at once—in parallel universes, so to speak. As time goes by, the number of configurations used by a non-deterministic Turing machine grows exponentially because each new action, combined with the previous actions, creates new branches of the computation tree. It is believed that a non-deterministic Turing machine could be exponentially faster than a deterministic one, although, as we will see, the question remains open.

For present purposes it is sufficient to know that the class NP coincides exactly with the class of decision problems whose solution can be verified efficiently in polynomial time. This alternative definition of the class NP is possible because a non-deterministic Turing machine solves these problems by "guessing" the solution, using its non-deterministic abilities to write on the tape many solutions in parallel and then checking if the solution is right. If that is the case, then the machine stops and accepts the input. All computations of the machine that do not lead to solutions fail to finish in the accept state and are ignored.

Some problems in the class NP are particularly hard in the sense that, if an efficient (i.e., polynomial time) solution for one of them exists, then an efficient solution for all the problems in NP must also exist. These problems are called *NP-hard.* (The traveling salesman problem is such a problem.) Decision problems that are NP-hard and are in NP are called *NP-complete.*

The traveling salesman problem, as formulated above, is not a decision problem, because it asks for the shortest path. However, it can be reformulated to become a related decision problem: Given a graph, is there a tour with a length smaller than a given number? The solution for the decision problem may be very hard to find (and is, in this case), but it is very easy to verify: Given a solution (a list of traversed cities, in order), verify, by simply summing all the inter-city distances in the path, if the total length of the trip is smaller than the given number.

Another class, P, is a subset of NP, and includes all decision problems that can be solved by a (deterministic) Turing machine in polynomial time. The decision problem associated with the shortest-path problem is in P: Is there a path between city A and city B shorter than a given number?

Stephen Cook (1971) showed that if an efficient algorithm is found for an NP-complete problem, that algorithm can be used to design an efficient algorithm for every problem in NP, including all the NP-complete ones. Richard Karp (1972) proved that 21 other problems were NP-complete, and the list of known NP-complete problems has been growing ever since. Interestingly, both Cook and Karp were at the University of California at Berkeley when they did the work cited above, but Cook was denied tenure in 1970, just a year before publishing his seminal paper. In the words of Richard Karp, "It is to our everlasting shame that we were unable to persuade the math department to give him tenure."

No one really knows whether an algorithm that solves NP-complete problems efficiently exists, but most scientists believe that it does not. If a polynomial time algorithm is discovered for one NP-complete problem (any problem will do), then that polynomial time algorithm can be adapted to solve efficiently (i.e., in polynomial time) any problem in NP, thereby showing NP is actually equal to P.

Whether P is equal to NP or different from NP is one of the most important open questions in computing and mathematics (perhaps the most important one); and it was also introduced in Cook's 1971 paper. It is, in fact, one of the seven Millennium Prize Problems put forth by the Clay Mathematics Institute in 2000. A prize of $1 million is offered for a correct solution to any of them.

Most scientists doubt that the P-vs.-NP problem will be solved any time soon, even though most believe that P is different from NP. If, as I find likely, there are problems in NP that are not in P, these problems will forever remain difficult to solve exactly and efficiently.

One problem known to be in NP but not necessarily in P is the problem of determining the factors of a composite number. A composite number is an integer that can be written as the product of two or more smaller integers. Given two numbers, it is very easy to verify if the first is a factor of the second by performing standard long division, which computers can do very, very rapidly.

Manindra Agrawal, Neeraj Kayal, and Nitinby Saxena (2004) have shown that it is also possible to determine, in polynomial time, whether a number is prime. However, if the factors of a very large integer are not known, finding them is very difficult. In such a case, the size of the description of the problem is the number of digits the number contains. All known algorithms for this problem take more than polynomial time, which makes it very difficult and time consuming to factorize numbers that have many digits. One may think this is a problem without practical relevance, but in fact the security of Internet communications rests largely on a particular cypher system: the RSA algorithm (Rivest, Shamir, and Adleman 1978), which is based on the difficulty of factorizing large composite numbers. Proving that P equals NP, were it to happen, would probably break the security of the RSA algorithm, which would then have to be replaced by some yet-unknown method.

Other computing paradigms, not necessarily equivalent to Turing machines, may move the border between tractable and intractable problems. Quantum computing, an experimental new approach to computing that uses the principle of superposition of quantum states to try a large number of solutions in parallel, can solve the factorization problem efficiently (Shor 1997). Working quantum computers that can deal with interesting problems, however, do not exist yet and aren't likely to become available in the next few decades. Very small quantum computers that are essentially proof-of-concept prototypes can work with only a few bits at a time and cannot be used to solve this problem or any problem of significant dimension. Many people, including me, doubt that practical quantum computers will ever exist. However, in view of the discussion of the evolution of technology in chapter 2 we will have to wait and see whether any super-Turing computing paradigms, based on machines strictly more powerful than Turing machines, ever come into existence.

Despite the fact that there are many NP-complete problems, for which no efficient algorithms are known, efficient algorithms have been developed for many fields, and most of us make use of such algorithms every day. These algorithms are used to send sounds and images over computer networks, to write our thoughts to computer disks, to search for things on the World Wide Web, to simulate the behaviors of electrical circuits, planes and automobiles, to predict the weather, and to perform many, many other tasks. Cell phones, televisions, and computers make constant use of algorithms, which are executed by their central processing units. In some cases, as in the example of the traveling salesman problem, it isn't feasible to find the very best solution, but it is possible to find so-called approximate solutions—that is, solutions that are close to the best.

Algorithms are, in fact, everywhere, and modern life would not be the same without them. In fact, a well-accepted thesis in computing, the Church-Turing thesis, is that any computable problem can be solved by an algorithm running in a computer. This may also be true for problems at which humans excel and computers don't, such as recognizing a face, understanding a spoken sentence, or driving a car. The fact that computers are not good at these things doesn't mean that only a human will ever excel at them. It may mean that the right algorithms for the job have not yet been found. On the other hand, it may mean that humans somehow perform computations that are not equivalent to those performed by a Turing machine.

The Church-Turing Thesis

Let us now tackle a more complex question: What can be effectively computed? This question has a long history, and many eminent thinkers and mathematicians have addressed it. Alan Turing, Alonzo Church, and Kurt Gödel were central to the development of the work that led to the answer. The Church-Turing thesis, now widely accepted by the scientific community even though it cannot be proved mathematically, results from work by Turing and Church published in 1936. The Church-Turing thesis states that every effectively computable function can be computed by a Turing machine. This is equivalent to the statement that, except for differences in efficiency, anything that can be computed can be computed by a universal Turing machine. Since there is no way to specify formally what it means to be "effectively computable" except by resorting to models known to be equivalent to the Turing model, the Church-Turing thesis can also be viewed as a definition of what "effective computability" means.

Technically, the concept of computability applies only to computations performed on integers. All existing computers perform their work on integers, which can be combined to represent good approximations of real numbers and other representations of physical reality. This leaves open the question of whether physical systems, working with physical quantities, can be more powerful than Turing machines and perform computations that no Turing machine can perform.

The Physical Church-Turing thesis, a stronger version of the Church-Turing thesis, states that every function that can be physically computed can be computed by a Turing machine. Whereas there is strong agreement the Church-Turing thesis is valid, the jury is still out on the validity of the Physical Church-Turing thesis. For example, it can be shown that a machine that performs computations with real numbers is more powerful than a Turing machine. Indeed, a number of computational models that might be strictly more powerful than Turing machines have been proposed. However, it remains unclear whether any of these models could be realized physically. Therefore, it isn't completely clear whether physical systems more powerful than Turing machines exist.

Whether or not the Physical Church-Turing thesis is true is directly relevant to an important open problem addressed in this book: Can the human brain compute non-computable functions? Do the physics of the human brain enable it to be more powerful than a Turing machine? More generally, can a physical system be more powerful than a Turing machine, because it uses some natural laws not considered or used in the Turing model?

In a 1961 article titled "Minds, Machines and Gödel," John Lucas argued that Gödel's first incompleteness theorem shows that the human mind is more powerful than any Turing machine can ever be. Gödel's theorem states that any mathematical system that is sufficiently powerful cannot be both consistent and complete. Gödel's famous result on the incompleteness of mathematical systems is based on the fact that, in any given formal system *S*, a Gödel sentence *G* stating "*G* cannot be proved within the system *S*" can be formulated in the language of *S*. Now, either this sentence is true and cannot be proved (thereby showing that *S* is incomplete) or else it can be proved within *S* and therefore *S* is inconsistent (because *G* states that it cannot be proved).

Lucas argued that a Turing machine cannot do anything more than manipulate symbols and that therefore it is equivalent to some formal system *S*. If we construct the Gödel sentence for *S*, it cannot be proved within the system, unless the system is inconsistent. However, a human—who is, Lucas assumes, consistent—can understand *S* and can "see" that the sentence is true, even though it cannot be proved within *S*. Therefore, humans cannot be equivalent to any formal system *S*, and the human mind is more powerful than any Turing machine.

There are a number of flaws in Lucas' theory, and many arguments have been presented against it. (One is that humans aren't necessarily consistent.) Furthermore, in order to "see" that a sentence is valid within *S*, the human has to be able to thoroughly understand *S*—something that is far from certain in the general case. Therefore, not many people give much weight to Lucas' argument that the human mind is strictly more powerful than a Turing machine.

Lucas' argument was revisited by Roger Penrose in his 1989 book *The Emperor's New Mind*. Using essentially the same arguments that Lucas used, Penrose argues that human consciousness is non-algorithmic, and therefore that the brain is not Turing equivalent. He hypothesizes that quantum mechanics plays an essential role in human consciousness and that the collapse of the quantum wave function in structures in the brain make it strictly more powerful than a Turing machine. Penrose's argument, which I will discuss further in chapter 10, is that the brain can harness the properties of quantum physical systems to perform computations no Turing machine can emulate.

So far no evidence, of any sort, has been found that the human brain uses quantum effects to perform computations that cannot be performed by a Turing machine. My own firm opinion is that the human brain is, in fact, for all practical purposes, Turing equivalent. I am not alone in this position. In fact, the large majority of the scientific community believes, either explicitly or implicitly, that the human brain is Turing equivalent, and that only technological limitations and limited knowledge stop us from recreating the workings of the brain in a computer.

Later in the book we will take a deeper look at the way the brain works. In the next few chapters, I will argue that the brain is effectively an immensely complicated analog computer performing computations that can be reproduced by a digital computer to any desirable degree of accuracy. Given a sufficiently precise description of a human brain, and of the inputs it receives, a digital computer could, in principle, emulate the brain to such a degree of accuracy that the result would be indistinguishable from the result one obtains from an actual brain. And if this is true for brains, it is also true for simpler physical systems, and, in particular, for individual cells and bodies. The idea that intelligent behavior can be the result of the operation of a computer is an old one, and is based, in large part, on the principle that the computations performed by human brains and by computers are in some sense equivalent. However, even if brains are, somehow, more powerful than computers, it may still be possible to design intelligent machines. Designing intelligent machines has, indeed, been the objective of a large community of researchers for more than fifty years. The search for intelligent machines, which already has a long history of promises, successes, and failures, is the subject of the next chapter.

5 The Quest for Intelligent Machines

Can a computer be intelligent? Can a program, running in a computer, behave intelligently in a human-like way? These are not simple questions, and answers have eluded scientists and thinkers for hundreds of years. Although the first mentions of non-human thinking machines can be found in Homer's *Iliad* (the automatic tripods of Hephaestus), and others can be found in a number of other literary works (among them Mary Shelley's *Frankenstein*, published in 1818), the question whether machines can exhibit intelligent behavior wasn't precisely addressed until later. Ada Lovelace seems to answer the question in the negative in one of her notes to Menabrea's "Sketch of the Analytical Engine Invented by Charles Babbage" (1843), stating that the Analytical Engine can do only whatever it is ordered to perform and has no pretensions to originate anything. Since the ability to create is usually viewed as one of the hallmarks of intelligence, this seems to point to the fact that Ada Lovelace believed that computers cannot become intelligent.

More recent researchers have been much more optimistic, sometimes overly so. Many predicted that intelligent machines would be available by the year 2000.

Artificial Intelligence

Modern Artificial Intelligence (AI) research began in the mid 1950s. A conference at Dartmouth College in the summer of 1956 led to great enthusiasm in the area. Many of those who attended that conference went on to become leaders in the field, among them Marvin Minsky, Herbert Simon, John McCarthy, and Allen Newell. AI laboratories were created at a number of major universities and institutes, including MIT, Carnegie-Mellon,

Berkeley, and Stanford. Groups of researchers began writing programs that solved many problems that previously had been thought to require intelligence. Newell and Simon's Logic Theorist (1956) could prove mathematical theorems, including some from Whitehead and Russell's *Principia Mathematica*, eventually finding demonstrations of dozens of theorems in Whitehead and Russell's masterwork—some of them more elegant than the ones known at the time. Arthur Samuel's checkers-playing program (1959) used alpha-beta search, a method to search the game tree, to play a reasonably good game of checkers and showed that it could even defeat its programmer. AI systems eventually became proficient at chess playing, action planning, scheduling, and other complex tasks.

However, intelligence is a more elusive concept than had once been thought. Though it is commonly accepted that intelligence is required in order for a human to address any of the problems mentioned in the preceding paragraph, it isn't at all clear that the techniques computers used to solve those problems endowed them with general human-like intelligence. In fact, those problems were tackled with specialized approaches that were, in general, very different from the approaches used by humans. For instance, chess-playing computers perform very extensive searches of future possible positions, using their immense speed to evaluate millions of positions per second—a strategy not likely to be used by a human champion. And computers use similarly specialized techniques when performing speech recognition and face recognition.

Early researchers tried to address the problem of artificial intelligence by building symbol-manipulation systems. The idea was to construct programs that, step by step, would mimic the behavior of human intelligence. They addressed and eventually managed to solve simple versions of problems that humans routinely solve, such as deduction, reasoning, planning, and scheduling. Each of those tasks, however, led researchers to unsuspected difficulties, as most of them are difficult to formulate and some of them are computationally hard. In fact, many problems we solve routinely in our daily lives are intractable, since they belong to the class of NP-hard problems. For instance, planning a sequence of actions, which can depend on each other, that takes the world from its present state to a desired state is a computationally difficult problem, known to be NP-hard. However, humans solve NP-hard problems every day without using inordinate amounts of time. This should not be viewed as evidence that the human brain is more powerful than a computer, only as evidence that heuristic approaches, which work most of the time and find approximate solutions, have been developed and incorporated into human reasoning by many millions of years of evolution.

Still, by developing sophisticated techniques, AI researchers have managed to solve many important problems, and the solutions are now used in many applications. For instance, train and airline schedules are commonly designed by AI-based systems, and many businesses apply techniques developed by AI researchers in data mining. Speech-recognition systems, which originated in the AI field, are now in wide use. In fact, AI techniques are already ubiquitous, and new applications are being found nearly every day.

However, the goal of designing a machine we can undoubtedly recognize as intelligent is still eluding AI researchers. One important reason for this is that intelligence is a slippery concept. Before chess-playing programs reached their present level, it was widely believed that playing chess at the championship level would require *strong AI*—that is, human-like artificial intelligence. However, when a specific procedure (an algorithm) was devised for playing chess, it became accepted that brute-force search and sophisticated position evaluation heuristics, rather than strong AI, could be used for that purpose. A similar change in our understanding of what AI is happened with many other problems, such as planning, speech understanding, face recognition, and theorem proving. In a way, we seem to attribute the quality of intelligence only to behaviors for which no algorithm is yet known, and thus to make intelligence an unreachable target for AI researchers.

However, there may be a more important reason why strong AI remains elusive. Humans behave in a way we deem intelligent because they interconnect knowledge and experiences from many different areas. Even the simple act of understanding speech requires a complete model of the world in one's mind. Only a program that stores a comprehensive model of the world, similar to the one used by humans, can behave in a way we will recognize as having human-like intelligence. Such a model is a very complicated piece of engineering. Humans were crafted by evolution to keep in their minds models of the world that are constantly used to disambiguate perceptions, to predict the results of actions, and to plan. We simply don't know how to explicitly build such a model. That explains why these earlier approaches to artificial intelligence yielded only very limited results.

How can we know whether a program is intelligent? If a machine can recognize and synthesize speech and can play a masterly game of chess, should it not be considered intelligent? How can we distinguish an intelligent machine from a machine that is simply running some specific algorithms on a given set of tasks? If a human being is able to talk to us in English and play a decent game of chess, we will certainly recognize him or her as intelligent, even if we had some doubts about his or her abilities in other domains. Are we not being too anthropocentric and too demanding of machines?

In addition to his major contributions to the theory of computing, Alan Turing also addressed the question of how to tell whether a machine is intelligent. In one of his first analyses, he assumed that a machine could eventually be made to play a reasonably good game of chess. He then wondered whether a human observer, with access only to the moves made on the board, could distinguish the machine's play from the play of a poor human player.

Eventually, Turing's ideas evolved toward what is now known as the Turing Test. In his seminal 1950 paper "Computing Machinery and Intelligence," Turing proposes to tackle the question "Can machines think?" Instead of trying to define elusive notions such as "intelligence" and "machine," he proposes to change the question to "Can machines do what we (as thinking entities) can do?"

Turing proposes a test inspired by the imitation game, a party game in which a man and a woman go into separate rooms and guests then try to determine which room the man is in and which room the woman is in by reading typewritten answers to questions asked of them. In the original game, one of the players attempts to trick the interrogators into making the wrong decision while the other player assists the interrogators in making the right one. Turing proposes to replace the woman with a machine and to have both the man and the machine try to convince the guests they are human. In a later proposal, Turing suggests that a jury ask questions of a human and a computer. The computer would pass the test if a significant proportion of the jury believed that it was the human.

The reason Turing's test has withstood the passage of time is that it avoids the most obvious anthropocentric biases we may include in the definition of intelligence. At the least, it prevents the jury from deciding whether something is intelligent or not by simply looking at its physical appearance.

On the other hand, the Turing Test still has a strong anthropocentric bias, because it forces the computer to imitate human behavior. For this reason, and also for other reasons, it remains a very difficult test for today's most advanced AI programs. To pass the test, a computer would have to possess human-like reasoning, memory, feelings, and emotions, since no limitations are imposed on what can be asked. Ultimately, any non-human-like behavior (such as the absence of emotion) can be used to distinguish the computer from the human. For instance, the jury can ask the program what is its oldest memory, or what was its most painful moment, or whether it likes sushi.

Turing, anticipating most of the objections that would be raised against his proposal, compiled a list of nine categories of objections. And indeed, in one way or another, all the arguments that came to be made against his test, and against the possibility of artificial intelligence, fall into one of the categories he listed. It is worthwhile to go through these objections in some detail. A few of them are easy to deal with, as they have no scientific basis and are either metaphysical or downright unreasonable.

The Theological Objection states that thinking is a result of man's immortal soul and therefore cannot be simulated by a machine. It is based on the assumption that humans are unique in the universe and are the only creatures with souls. This objection, which is based on the duality of mind and body, will be analyzed in more detail in chapter 10.

The Informality of Behavior objection is based on the idea that there is no set of rules that describes what a human will do in every possible set of circumstances, which implies that human behavior can never be simulated by a computer. In a way, this argument is equivalent to the idea that human intelligence is non-algorithmic, which makes it very similar to one of the other arguments presented below.

The Heads in the Sand objection states that the consequences of a thinking machine would be so dreadful that one will never arise, presumably because humanity would steer away from its development. But humanity doesn't seem to have been able to steer away from any technology in order to avoid the risks it presents. This argument, however, is related with the dangers of super-intelligences, addressed briefly in chapter 12.

Two objections are based on the argument that the brain is not Turing equivalent, either because it can (in some unspecified way) compute non-computable functions or because the inherently ability of brain cells to work with real-valued signals gives it additional power. This objection will be analyzed in chapters 8 and 9 when we examine the workings of the human brain and the questions raised by our efforts to emulate it. However, it must be said that many people are strong believers in this argument, and it must be conceded that the question remains largely unresolved.

Another objection, first advanced by Ada Lovelace, is based on the argument that computers necessarily follow fixed rules and therefore are incapable of originality, and that their behavior always leads to predictable results. This objection ignores the fact that very complex systems, even if completely defined by fixed rules, have utterly unpredictable behaviors, as modern engineers and scientists are well aware.

Another objection is based on the idea that humans have extra-sensory perception, and that it cannot be emulated by a machine. Because extra-sensory perception remains to be observed under controlled conditions, this objection carries little weight.

A final philosophical objection, which may be more profound than the others, argues that intelligence can only originate in consciousness, and that a symbol-manipulating machine can never attain consciousness. The argument that a computer can never exhibit consciousness would be made much later, more explicitly, in John Searle's (1980) Chinese Room thought experiment.

Searle's thought experiment begins with a specific hypothesis: Suppose that AI researchers have succeeded in constructing a computer that behaves as if it understands Chinese. The computer accepts Chinese characters as input and, by following a set of fixed rules, produces Chinese characters as output. Suppose, Searle argues, that this computer performs its task so well that it passes the Turing Test, fooling a Chinese-speaking jury. By the criterion of the Turing Test, we would then conclude that the computer understands Chinese. Searle, however, argues against that conclusion. He asks the reader to suppose that he is in a closed room and that he has paper, pencils, erasers, and a book containing an English-language version of the computer program. He can receive Chinese characters, process them by looking at the instructions in the book (thereby simulating the behavior of the program), and produce Chinese characters as output. Because the room-operator system can pass the test without understanding a word of Chinese, we must infer that the computer doesn't understand Chinese either.

Searle's argument hides, of course, a strong anthropocentric bias. Searle somehow assumes that a system (the room, plus the computer instructions, plus the computer, either human or digital) can, magically, conduct a conversation in Chinese without understanding a word of Chinese. Searle would attribute the magic quality of understanding only if a human being, somewhere in the loop, was able to somehow change a blind computational process into a conscious process. This is exactly the bias Turing tried to avoid by proposing his blind test, wherein the jury cannot be influenced by knowledge as to whether or not a human is involved.

Turing specified how the interaction between the jury and the players should take place. The questions and the answers should be typewritten, to avoid difficulties with the understanding of spoken language that are not deemed central to the problem. Thanks to recent technological advances in speech recognition and speech synthesis, this particular difficulty might become irrelevant in the near future, and we could imagine a full-fledged Turing Test in which live conversation with a human and with a synthesized avatar would replace the typewriter interface that Turing specified.

A number of experiments have shown that care must be taken to avoid oversimplification of the requirements expressed by Turing. A simple program called Eliza, written in 1966, used simple rules to mimic the behavior of a psychotherapist in such a way that it convinced many people it was a real psychotherapist. A program called Parry, written in 1992, mimicked the behavior of a paranoid schizophrenic convincingly enough to fool approximately half of the psychiatrists who were shown the dialogues. Over the years, many other programs have been said to have passed the Turing Test. However, all of them succeeded only on very restricted versions of the test, and it is understood with significant confidence that decades will elapse before a computer passes an unrestricted version of the Turing Test. However, this doesn't mean that it will never happen.

In 1990, the fortieth anniversary of the first publication of Turing's paper "Computing Machinery and Intelligence," a colloquium at the University of Sussex brought together a large group of academics and researchers on the subject of the Turing Test. One result of this colloquium was the creation of the Loebner Prize, awarded to the program considered by the judges to be the most human-like. The present format of the competition is that of a standard Turing Test, including the typewritten answers. In the 2008 edition, the program Elbot fooled three of the twelve judges into thinking the computer was the human after a five-minute conversation. That came tantalizingly close to the original objective of fooling a "significant fraction" of the judges, proposed as 30 percent by Turing. But a five-minute conversation doesn't amount to much. More recent Loebner Prize competitions have specified longer and more demanding tests. The winner of the 2013 Loebner Prize, the Mitsuku chatbot, is, at the time of writing, available for interaction on the Web. If you test it, you will soon understand that it is indeed a program, that it has a fairly superficial model of the world, and that it doesn't come close to passing what Turing intended to be an unbiased test for true human-like intelligence.

Although the Turing Test remains the most independent and unbiased test ever proposed to assess machine intelligence, it has weaknesses, and many people have objected to its use. The most obvious objection is that the power of the test depends on the sophistication of the jury. Some people may be easily fooled, but others, more knowledgeable about the technology and the issues under discussion, may be much harder to be led astray. A more serious objection is that the test may classify intelligent behavior as non-intelligent simply because it isn't close enough to human behavior. It is entirely possible that even a super-human intelligence such as the mythical computer HAL in Arthur C. Clarke's 1968 novel *2001: A Space Odyssey* would fail to pass a Turing Test because it would think in ways too different from the ways humans think.

Overall, the Turing Test has not been central to the development of artificial intelligence. AI researchers have solved many specific problems of immediate practical use, and have not dedicated significant effort to the development of programs with human-like behavior. Nonetheless, the Turing Test remains significant in the history of computing and artificial intelligence for the philosophical questions and challenges it raises.

Learning from the Past

Research in artificial intelligence is concerned with the way humans address and solve problems. Although it is difficult to define exactly what specific areas are included in the field of artificial intelligence, all of them are, in one way or another, related to the ways humans exhibit intelligent behavior. Human intelligence gives us the abilities to plan, to understand language, to interpret the world, and to control our movements in order to move around and to manipulate objects. These abilities correspond, largely, to AI's sub-fields of planning, natural-language processing, artificial vision, and robotics. Most problems in these areas have been studied for decades by a multitude of approaches, but they have proved to be, in general, much harder than they seemed at first. For example, the problem of creating a description of the objects in one's field of vision, once given as a summer project to students at MIT by Seymour Papert (1966), remains unsolved after the efforts of thousands of researchers. Only very recently have developments in this area managed to create systems that are able to address that problem with some success.

There are many works on the subject of artificial intelligence and on many of its subfields, among them the reference textbook by Russell and Norvig (2009). AI is such a large field that I cannot even begin to address here the many techniques and theories that have been developed over the years by researchers.

There is, however, one subfield of AI so important to the problems we are discussing that it deserves a special place in this book. That subfield, machine learning, is central to all activities requiring intelligence and is concerned with the ability of machines to learn from experience—the most substantial ability that a system must possess in order to exhibit intelligent behavior. Of the many excellent books on the subject I strongly recommend Friedman, Hastie, and Tibshirani 2001, Vapnik 1998, Domingos 2015, and Michalski, Carbonell, and Mitchell 2013.

Human learning takes many forms, ranging from those that enable us to learn how to read and write to those (more innate and primitive) that enable us to learn a spoken language, to recognize objects, or even to walk. Machine learning is concerned with the replication of these and other related abilities in programs. Natural-language processing, artificial vision, robotics, and planning, all important fields of AI, would not have developed without the techniques invented by researchers in machine learning.

The recognition that learning is essential for the development of AI systems wasn't always as clear as today. Early researchers (Laird, Newell, and Rosenbloom 1987; Rich 1983) tried to build into programs fully formed models of the world that would enable these programs to conduct a conversation or to process an image. The symbolic approach to artificial intelligence (which became known as Good Old-Fashioned Artificial Intelligence, abbreviated GOFAI) didn't work as well as had been expected, because the problem of creating an artificially intelligent system was much harder than had been expected. Building an explicit and complete model of the world, for instance, is now recognized to be a very, very difficult task, because no human being can explicitly describe such a model, much less write it down in a form that a computer can use. Every one of us has his or her own model of the world—a complex, fluid, multi-faceted model that differs from the models of others in many ways. Furthermore, adapting symbolic systems to the uncertainty always present in real-world problems proved to be much more difficult than had been expected. Object recognition, for instance, is influenced by camera noise and lighting conditions, phenomena that are hard to incorporate in explicit logic models.

Eventually most researchers turned to the idea that an AI system must have a built-in capacity to adapt and to learn from experience—a capacity that would, in time, create the world models that are needed. This implies that what must be created is not a system intelligent from the very beginning, by design, but a system that can learn to be intelligent by adapting its own internal structures on the basis of experience. This idea can be traced back, again, to Alan Turing, who, in the same article in which he proposed the Turing Test, suggested that it might be easier to program a machine that learns than to program a fully operational thinking machine:

> Instead of trying to produce a programme to simulate the adult mind, why not rather try to produce one which simulates the child's? If this were then subjected to an appropriate course of education one would obtain the adult brain. Presumably the child brain is something like a notebook as one buys it from the stationer's. Rather little mechanism, and lots of blank sheets. Our hope is that there is so little mechanism in the child brain that something like it can be easily programmed. (Turing 1950)

Learning is, in fact, what the human brain does, as do the brains of many other animals, to a lesser extent. The brain of a newborn human doesn't enable it to perform many actions. A newborn cannot walk or talk, has only basic visual perceptions and motor skills, and would not survive alone in the wild. Yet, as the years pass, the structures in the brain adapt in such a way that the child learns to walk, to talk, to move, and to interact with the world in a meaningful way, becoming an autonomous and self-sufficient agent.

What is true of humans is also true, to a lesser degree, of non-human primates, other mammals, and even non-mammals. Their brains do not come fully formed. Instead, they adapt in such a way as to be able to endow their owners with whatever skills they need to survive. What is unique to humans is the extension of the learning process, which takes many more years than in any other species and which transforms a defenseless newborn into a fully autonomous adult. The extension of our ability to learn is, therefore, the crucial characteristic that separates our brains from the brains of other animals. Reproducing this ability in a program became a major field in AI research: the field of machine learning.

The central idea of machine learning is relatively easy to explain, as are some of the machine learning techniques that have been developed and successfully used in many domains. The quintessential problem of learning consists, in a very simplified way, of inferring general rules or behaviors from a number of specific concrete experiences. This is called *inductive learning*, and it can take many forms. Inductive learning can be applied to many problems and has been a topic of countless scientific articles, books, and dissertations, including my own dissertation (Oliveira 1994).

In its simpler form, inductive learning is performed by learning a general rule from a set of labeled instances. In this simple case, each instance is described by the values of a fixed set of attributes and by a label that specifies the class of the instance. Consider, for example, the hypothetical problem of learning, from experience, whether a specific day is a good day on which to play tennis. You remember some days in the past weren't good days on which to play, and you registered some weather variables for those days in a table that may look similar to table 5.1. To avoid the negative experience that stems from trying to play tennis on a day not good for the sport, you decide to infer, from this table, a rule that helps you decide whether a day with some specific characteristics is a good day on which to play tennis. (This is, of course, a contrived and artificial problem, but it serves to illustrate the central idea. You can imagine that, by performing some analysis of the table, you could infer a rule that could be used to classify days as either good or not good for playing tennis.)

Given a small number of instances, there are many possible solutions, or rules, but some may seem more reasonable than others. In its simplest

Table 5.1
Learning instances.

Temperature	Humidity	Wind	Play tennis?
70	95	5	Yes
32	80	10	No
65	80	20	No
75	85	10	Yes
30	35	8	No
75	35	8	Yes
72	35	25	No

form, machine learning aims at finding the classifier with better expected performance in future instances not yet observed or experienced. In this context, a classifier is simply a rule or set of rules that, when given the values of the attributes of one instance, produces the label of that instance. In our example, an instance corresponds to a specific day, attributes of the instance are the characteristics of that day, and the objective is to label each day with either Yes or No.

Visualizing the instances may sometimes provide hints. For instance, by plotting the instances in table 5.1, using only the temperature and the wind values, one obtains the graphic shown in figure 5.1, which seems to show that days that are good for playing tennis cluster together in a two-dimensional space. In fact, this visualization leads immediately to one simple approach to classifying days: When classifying a new instance, simply look in the table for the instance that is more similar than all others, in terms of attributes, to the instance under analysis, and use the label of that instance as the answer. For example, a day with the attributes (Temp, Humidity, Wind) = (34, 80, 12) would be very similar to the day in the second row of the table and, therefore, would not be a good day on which to play. This useful algorithm, called "nearest neighbor," led to a large family of methods by which to perform induction, which are sometimes grouped together as *similarity-based learning*. However, there a number of difficulties that make the nearest neighbor method not applicable in many cases. In particular, similarity-based methods require a good measure of similarity between instances (something that isn't always easy to derive), and they may be inefficient if they have to search extensive databases to obtain the label of the nearest instance.

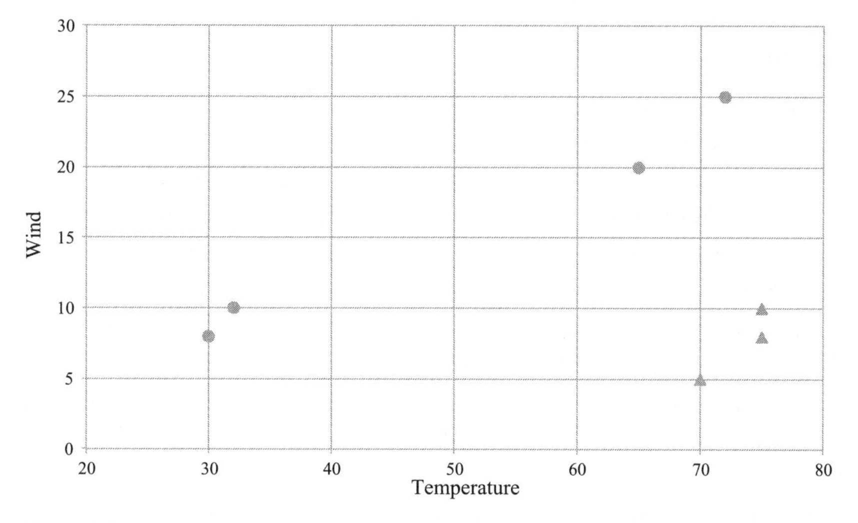

Figure 5.1
A graphical depiction of table 5.1, using the temperature and wind attributes. Triangles represent days that are good for playing tennis and circles represent the other days.

We now know that the problem of inferring a general rule from specific examples is not well posed, since it always admits many possible solutions and there is no universal way to select the best one. In this particular case, it may happen that you enjoyed playing tennis on days that were warm and not windy, but it may also happen that some other particular combination of characteristics was responsible for your enjoyment of the sport. Furthermore, it may happen that what made you enjoy playing tennis in the past will not have the same effect in the future.

David Hume (1748) was the first to put the finger on the central problem of induction by arguing there is no obvious reason why the future should resemble the past. There must be a belief in some regularity that enables us to make inferences from a finite set of instances. Hume saw clearly that we can never be sure that the data we obtain about a specific phenomenon suffice to enable us to predict the future. In Hume's opinion, even things as well known as "the sun rises every day" cannot be used as predictors of the future, whereas mathematical truths obtained by deduction are demonstrably true:

That the sun will not rise tomorrow is no less intelligible a proposition, and implies no more contradiction, than the affirmation, that it will rise. We should in vain, therefore, attempt to demonstrate its falsehood. Were it demonstratively false, it would imply a contradiction, and could never be distinctly conceived by the mind. (Hume 1748)

Hume's basic point, that induction from past experience cannot provide guaranteed results, remains as valid today as it was in the eighteenth century. Our ability to learn from experience exists only because there is some regularity in the data that, for some reason, we are able to explore. This is called the *inductive bias*, and deciding what it should be has occupied researchers for decades, leading to a variety of learning algorithms from many different schools of thought. Pedro Domingos, in his 2015 book *The Master Algorithm*, systematizes the approaches and biases that have been developed by machine learning researchers into five large families of methods, each based on a specific idea. Some of these methods are briefly described in this chapter.

It is possible to demonstrate that all learning algorithms have the same performance when their respective performances are averaged over all possible problems in a given domain. This is called the *No Free Lunch Theorem*; it is equivalent to the statement that all learning algorithms are equally good if a preference for a specific learning bias can't be established (Wolpert 1996). For real problems, it is usually possible to establish a preference for some explanations over alternative ones. That makes it possible to learn from experience, as we do every day.

Of course, the problems can be much harder than the one in the example cited above. They can, in fact, be arbitrarily hard, because almost any perceptual human experience can be framed as an induction problem, specified by a table of instances, with some minor adaptations. Consider, for example, the extremely hard problem of recognizing an object in an image, a problem that remains essentially unsolved even after many decades of research. This is a somewhat simplified version of the problem of scene recognition to which I referred earlier in this chapter. In this simpler problem, one wants a classifier that recognizes when a particular object (for example, a car) is present in a picture, and answers with a Yes or a No the question "Is there a car in this picture?" The picture itself can be an array of gray-scale pixels in which a pixel can take any value from 0 to 1, where 0 means black and 1 means white. Color images are similar, but in these images each pixel requires three values to specify the exact color and its intensity.

This problem can be framed very easily as a typical induction problem, where a rule is to be learned from a set of instances. In this case, the attributes are the pixels in the image. The label is a Yes if there is a car in the picture and a No if there is no car. The table would necessarily be a much larger table. If one uses one megapixel images, there are a million attributes, each corresponding to one column in the table. The table includes an additional column for the label that contains, in this case, a Yes or a No. Despite the superficial similarity of these two problems and the fact that they can be formulated directly as inductive learning problems by presenting the table of instances, the solutions for them are, because of their intrinsic complexity, profoundly different.

For the problem in table 5.1, it is relatively easy to imagine that there is a simple rule that differentiates between days good for playing tennis and days not good for playing tennis. Perhaps some simple rule stating that warm days with little wind are good will work well. Computers can deal with such tasks very easily, and in fact machine learning is routinely used to solve them. Machine learning is used, for instance, to analyze your transactions when you use a credit card or when you make a succession of withdrawals from your bank account. The rules used are probably derived from many millions of previous transactions. Given the patterns observed in your most recent set of transactions, they specify whether an alarm should be raised and the transactions should be inspected. Computers deal with such comparatively simple problems by learning sets of rules that work not only in the cases used during training but also in new cases that have never appeared before. This is exactly the point of machine learning: deriving rules that can be applied in instances that have not been seen before by the system but that are, in some sense, similar to the ones that have been seen.

Object recognition is conceptually similar but, in practice, much harder. Although humans learn to recognize cars, and can easily tell when a car is present in a picture, we don't know exactly what mechanisms the brain uses to perform this task. Presumably our brains build internal representations of cars as we become familiar with them. We learn to recognize parts of cars (wheels, doors, headlights, and so on), and somehow our brain puts these things together so that we can recognize a car. Of course, even a wheel or a door is hard to recognize, but presumably, it is somewhat easier to recognize than a whole car.

The field of machine learning has evolved so much that the recognition of objects is now within reach of computer programs. There are many algorithms that can be used to perform inductive learning; I will cover only a few of them here. Learning from patterns is very relevant to the central topic of this book, since it is believed that much of the human brain's plasticity results from related mechanisms. Plasticity, the ability of the brain to change and adapt to new inputs, is believed to be the central mechanism involved in learning in animals and in humans.

Given a set of instances, each one with a label, the objective is to derive some rule that distinguishes the positive instances from the negative ones. More general cases in which there are several possible label values can be handled by the same basic techniques. One way to derive a rule is to infer logical or mathematical formulas that can be used to distinguish the positive instances from the negative ones. For instance, one may imagine simple rules of the form "If it is a warm day and it is not windy, then it is a good day for playing tennis" or more complex rules represented by logical combinations of these simple primitive rules.

One very successful class of learning algorithms, inference of decision trees (Quinlan 1986; Breiman et al. 1984), formulates rules as a tree, which is then used to classify previously unseen instances. Figure 5.2 shows an

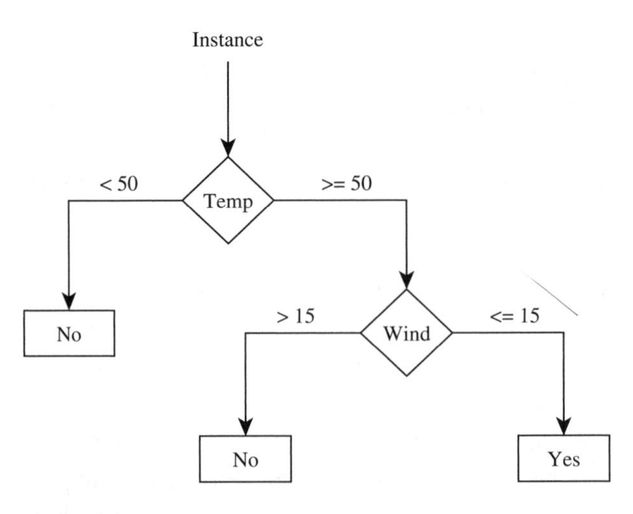

Figure 5.2
A decision tree for the tennis problem.

example of a decision tree that could be used to classify the instances of the tennis problem.

Decision-tree learners try to infer the best possible tree, typically preferring simple trees over more complex ones on the basis of the argument (first put forth by William of Occam, in the fourteenth century, and commonly known as Occam's Razor) that simpler explanations that fit the available data have more predictive power than more complicated explanations. The statement that entities must not be multiplied beyond necessity, attributed to Occam by John Punch (Crombie 1959), is usually interpreted as a defense of simpler theories versus more complicated ones.

The simple and intuitive idea behind Occam's Razor led to many complex discussions about what "simple" means, and why simpler explanations should have more predictive power. This bias in favor of simplicity has been extensively studied and debated by philosophers, mathematicians, and computer scientists (Solomonoff 1964; Blumer et al. 1987; Domingos 1999). It is possible to conclude that simpler rules have, in general, higher predictive power, although the definition of "simpler" has to be formulated in very explicit mathematical terms.

To obtain simpler explanations, many decision-tree learners proceed by greedily selecting the attributes to be tested on the basis of how much information they bring about the label. Attributes that are more informative (i.e., more highly correlated with the label) are chosen first, thus creating a preference for smaller and simpler trees.

Decision trees can be used very effectively in a number of domains because they are a structured and understandable way to represent a decision procedure based on a set of tests. For these reasons, they have been extensively used in medical diagnosis (Kononenko 1993), in financial analysis (Magee 1964), and in many other areas of science and industry. However, they are not the best tool for performing inductive inference in many problems, particularly when the attributes are numerous and real-valued and the boundaries of the classes are complex and high-dimensional. Other, more sophisticated classification methods can be used for such problems.

Perceptrons and Artificial Neural Networks

The use of decision trees is just one of the many techniques used to infer (or learn) general classification rules from a set of instances. These rules can take many forms and can be inferred in many different ways. One way, for instance, is to convert the attributes into numerical values (if they are not already numerical) and then perform arithmetic computations on them. One obvious and popular approach is to compute a weighted sum of the input attributes, with the weights determined by some criterion, and then compare the weighted sum with a threshold. The result of this comparison defines the class of the instance. In this way, one is weighting the different inputs, in much the same way one would ponder the different variables that influence some difficult decision. If the weights are chosen carefully, the value of the sum can be used to separate the instances according to the values of their class labels. Other, more complex forms of combining the input values can also be used.

The idea of computing a weighted sum of the values of the attributes is particularly appealing, in part because of its simplicity and in part because it is vaguely inspired by the way neurons in the brain work. Warren McCulloch and Walter Pitts (1943) were the first to propose that neurons in the brain perform a weighted sum of their inputs, computing threshold functions whose outputs are then combined to perform the complex tasks involved in thinking. Frank Rosenblatt (1958) proposed the *perceptron*, a simplified model for the neurons in the brain. Real neurons do indeed perform computations on their inputs that can be viewed, in a very simplified way, as weighted sums of the values of these inputs. Rosenblatt's perceptron model, which was originally designed to be implemented as hardware, is similar to an artificial neuron, since it computes a weighed sum of its inputs and compares it with a given threshold. If the value is higher than the threshold, the perceptron "fires" and its output is 1. Otherwise, the output is 0. The computation performed by a simple perceptron, with four inputs (x_1, x_2, x_3, x_4) and four weights (w_1, w_2, w_3, w_4), is illustrated in figure 5.3. The additional weight w_0 is connected to a constant input with value 1 and defines, in reality, the value of the threshold.

Perceptrons drew significant interest for two main reasons. The first was that they were, in some ways, similar to actual neurons in their generic behavior. Real neurons also fire only when the total amount of excitation they receive is high, and remain "silent" when the total amount of excitation is low. The computation performed by a real neuron is much more complicated than a simple sum and, in the majority of the cases, is not likely to correspond to a linear sum of the inputs. However, if one wants to

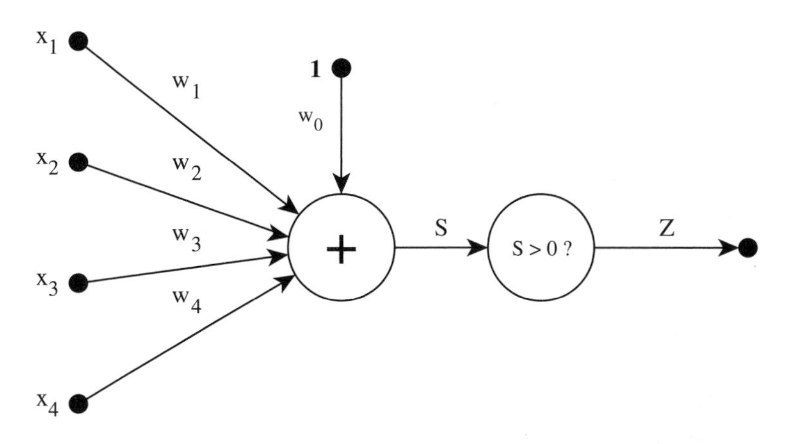

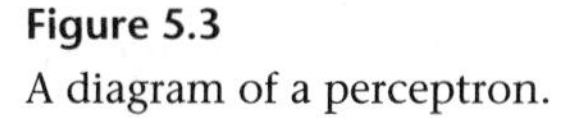
Figure 5.3
A diagram of a perceptron.

define a very simplified mathematical model of the behavior of a single biological neuron, it is reasonable to use, to a first approximation, the perceptron model.

The second reason why perceptrons caused some excitement when first proposed was that a learning algorithm to train perceptrons was available and could be used to derive the values of the weights and the threshold. This algorithm corresponds to a mathematical rule that adapts a perceptron's input weights in order to make the perceptron perform the desired task. This rule is very simple. Every time an instance is presented, the rule changes each weight if the desired output, Y, is not the same as the actual perceptron output, Z. The weight is changed by a small amount, controlled by a parameter α, in such a way that it moves the value of the weighted sum (before the threshold is applied) in the direction of the desired value, in accordance with the equation

$$w_i(t+1) \leftarrow w_i(t) + \alpha(Y - Z)x_i.$$

When input x_i is positive, this equation simply states that at time $t + 1$ weight w_i is increased if the desired output Y is higher than the observed output Z, and is decreased otherwise. If x_i is negative, the changes occur in the opposite direction. The change defined by this equation implies that the weighted sum, for this particular input pattern, will now be slightly larger if the desired output is larger than the observed output, and will be slightly smaller if the opposite condition is true. This update rule is applied to every weight w_i. If some fairly simple conditions are met, this rule is guaranteed to work and to eventually lead to a set of weights that makes the perceptron classify each and every input correctly.

The perceptron model appeared to be a significant step toward the emergence of self-configurable systems that would learn by themselves to perform any desired task. Perceptron-based learning was seen, at the time, as a very promising avenue to develop adaptive brain-like behavior in computers and electronic systems. In fact, in 1958 the *New York Times* optimistically called the perceptron "the embryo of an electronic computer that [the Navy] expects will be able to walk, talk, see, write, reproduce itself and be conscious of its existence" (Olazaran 1996).

Alas, things turned out not to be so straightforward. The challenge lies in one of the conditions that has to be met in order for the perceptron update rule to work and in the fact that the rule is applicable only to single perceptrons. Since the perceptron computes a simple weighted sum of the inputs, it computes what is mathematically known as a linear function, and it can distinguish only things that are linearly separable. For instance, if the instances are defined by only two real-valued attributes, x_1 and x_2, they can be visualized as points in two-dimensional real space, one dimension for each attribute. In this simple case, a perceptron can separate the positive and negative instances only if they can be divided by a straight line. Classes with more complex separations cannot be learned by a perceptron, as Minsky and Papert made abundantly clear in their 1969 book *Perceptrons*. That book dampened the enthusiasm for perceptron-based approaches that existed at the time and led researchers to look for new ideas for the development of intelligent machines.

Figure 5.4 illustrates two examples of hypothetical problems with two real-valued attributes. When there are only two attributes, the problems can be conveniently drawn in two dimensions, since every instance corresponds to one point in the two dimensional plane. The problem illustrated in figure 5.4a can be learned by a perceptron, because a straight line can separate the positive instances (filled circles) from the negative instances (non-filled circles). The problem illustrated in figure 5.4b cannot be learned by a perceptron, because there is no straight line that can be used to separate the two types of instances. The positive instances of the problem in table 5.1 (represented in figure 5.1 using only two of its three dimensions) can also be separated from the negative instances by a straight line. This problem can therefore be learned by a perceptron.

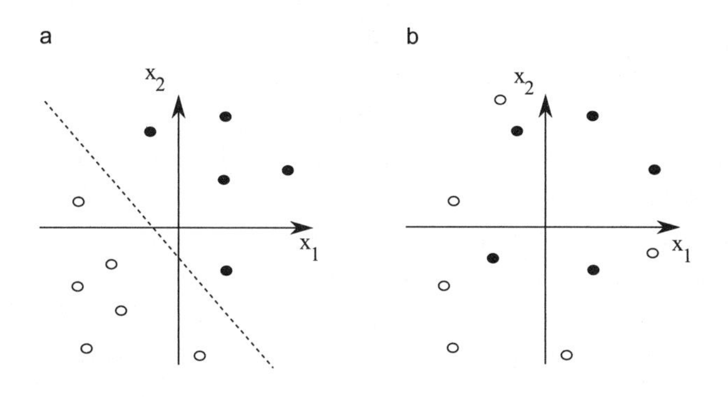

Figure 5.4

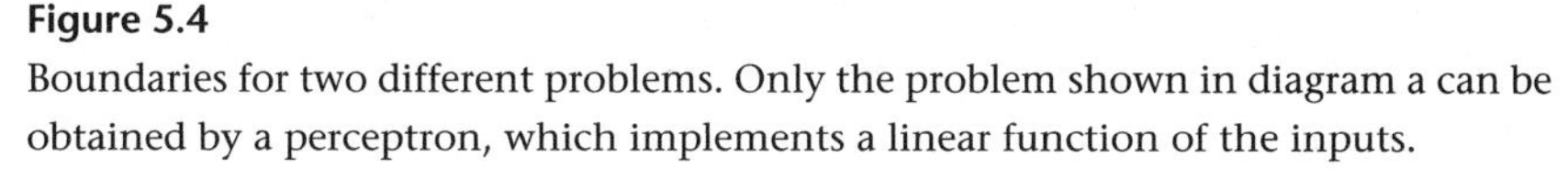
Boundaries for two different problems. Only the problem shown in diagram a can be obtained by a perceptron, which implements a linear function of the inputs.

The negative result that single perceptrons cannot learn more complex problems would not be, by itself, a severe blow to the perceptron-based approach, because, after all, one perceptron is only the model for a single neuron. No one expects a single neuron to perform tasks requiring intelligence, and indeed a number of results have shown that multi-layer artificial neural networks can approximate arbitrarily well any desirable behavior and can, therefore, be used to perform difficult tasks if the weights are assigned correctly (Hornik, Stinchcombe, and White 1989). Artificial neural networks (often referred to simply as neural networks) are computational models based on the interconnection of simple neuron models. They don't intend to reproduce faithfully the behavior of real neurons, but they represent a very useful abstraction.

It was once hoped that some learning rule could be used to train networks of perceptrons arranged in multiple layers. Such a learning rule would have had to solve the so-called *credit assignment problem*, which consists in determining the values that should be assigned to the weights in order for the whole network to perform the desired task.

The credit assignment problem was shown to be very difficult to solve. We now know that this problem is NP-hard. A number of results have shown that training even simple networks of perceptrons is an intractable problem (Blumer and Rivest 1988), implying that there is no known efficient algorithm, guaranteed to work in all cases, that will assign the right weights to a multi-layer neural network.

However, the hope for an efficient training algorithm came to be realized, partially, when algorithms that could be used to train soft-threshold multi-layer perceptrons were rediscovered and were popularized, mainly through the work of David Rumelhart, James McClelland, and the PDP Research Group (1986). The breakthrough was made possible by the realization that if standard perceptrons, with their hard thresholds, were to be replaced by soft-threshold perceptrons, a training algorithm could be devised that would, at least in principle, derive the required weights for an arbitrary network of perceptrons. This result was even extended to networks of perceptrons with feedback, in which the outputs of perceptrons were fed back onto themselves, directly or indirectly (Werbos 1988; Almeida 1989).

This algorithm opened the way to an approach called *connectionism*, which is still actively being used even though it has progressed more slowly than its supporters expected at first. Connectionists believe that networks of very simple processing units, with the right configuration and the right weights, can be used to perform very complex tasks, and that eventually they may exhibit intelligent behavior.

When a network of soft-threshold perceptrons (such as the one illustrated in figure 5.5) is used, the output neuron computes a function that is smooth and well behaved. In mathematical terms, this means that small changes in the inputs or in the connection weights lead to small changes in the output. In the example illustrated in figure 5.5, there are five input units, three hidden units (between the input and the output), and one

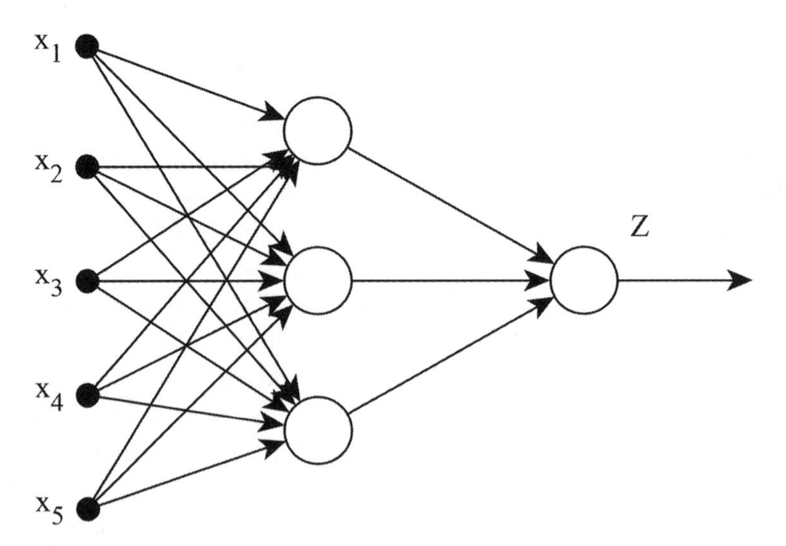

Figure 5.5
A diagram of a multi-layer perceptron with an input layer, a middle layer, and a output layer.

output unit. Each connection has a weight (not shown) that can be adapted to make the multi-layer perceptron perform the desired task.

Mathematically, it is possible to compute, for any possible input pattern, how the value of the output neuron varies with a small variation of each interconnecting weight. Therefore, it is easy to determine how a small change in any weight in the network affects the output. This makes it possible to compute how the weights should be changed in order to make the value of the output neuron move in the desired direction.

More precisely, what is computed is the derivative of the output error with respect to each of the weights. This collection of derivatives is called the *gradient*, and, because the algorithm changes the weights in order to minimize the error, the technique is called *gradient descent*. A simple mathematical formulation of this computation, called *back-propagation* (Rumelhart, Hinton, and Williams 1986), starts by computing the output error and back-propagates it until the input neurons are reached, obtaining, in the process, the complete gradient of the error with respect to the weights. That gradient can then be used to adjust the weights in order to reduce the output error.

Conceptually, back-propagation solves the credit assignment problem, thereby showing an effective algorithm for training networks of perceptrons exists—something that seems to go against the theoretical limitation that training multi-layer perceptrons is computationally hard. The apparent contradiction is explained by the fact that back-propagation can indeed be used to train complex networks but will, in many cases, fail to find a good solution (that is, one in which the error is small and the rule learned by the network is useful).

Some problems, such as recognizing a car in an image or recognizing a spoken sentence, are simply too complex to be solved directly by the application of back-propagation. Such problems require significant transformations of the input values in order to derive complex internal representations that can then be combined with previous knowledge to yield the final answer. In these cases, back-propagation may be too slow or may generate inappropriate weights, thereby failing to solve the credit assignment problem. This is a consequence of the fact that the error function, whose minimum is sought by the back-propagation algorithm, is simply too hard to minimize.

Back-propagation is only one of the many different techniques that have been invented to adjust the weights that interconnect these simple, neuron-like units, called perceptrons. Many other approaches and models have been proposed, making this field one of the most complex and well researched in the larger area of machine learning. Recently, new developments, called collectively *deep learning*, have extended significantly the performance and range of applicability of neural networks. Deep learning methods use large amounts of training data, sophisticated multi-layer architectures, and various neuron models to derive complex structures that include intermediate representations of complex attributes. Some deep learning techniques are inspired not only by standard connectionist tools (such as back-propagation) but also by advances in neuroscience and, in particular, by the patterns observed in actual nervous systems.

Deep learning techniques have been applied to computer vision, to automatic speech recognition, to game playing, and to natural-language processing. Neural networks trained using deep learning have been shown to produce state-of-the-art results that represent significant advances in the field. Ultimately, however, it is not likely that existing machine learning techniques, including deep learning, can be used to autonomously derive the complete set of connections in a neural network that has billions of artificial neurons and is comparable in complexity to the human brain. They will certainly be used to perform complex tasks that required, until now, the intervention of humans, such as driving a car or recognizing a face. However, more significant advances will be required if the objective is to design systems that, by learning from experience, will end up exhibiting human-like intelligent behavior.

The Formula of Reverend Bayes

The approaches to learning from examples described above may seem somewhat *ad hoc* in the sense that a given framework must be selected, and only then are its parameters adjusted to the available data. Before you actually perform any learning, you have to select the model that will be used, and only then is the model inferred from existing data. You may, for instance, select decision trees, or neural networks, or one of the hundreds of different ways to construct classifiers. Only after this initial decision, which seems rather arbitrary, can you apply the learning algorithms to derive the structure and parameters of the classifiers. The theory behind such an approach doesn't seem very solid, since at the beginning of the process lies a very arbitrary choice of the learning paradigm.

In practice, one would select a few different methods, try them all, and then pick the one that provides the best results. This improves the process somewhat, but it still leaves much to be desired, since the initial choice of the classifiers that will be tested remains rather arbitrary and doesn't follow any clear rules.

Is there a better way? Is there a mathematical formulation that doesn't involve arbitrary choices and that provides the best answer, obtaining the best possible classification for all future instances? The answer is, somewhat surprisingly, "Yes and no." Thomas Bayes was the first to discover the answer to this question. He presented it in an essay that was read to the Royal Society in 1763, two years after his death. The presentation, made by Richard Price, was titled "An Essay towards solving a Problem in the Doctrine of Chances." It described a somewhat restricted version of what is now known as Bayes' Theorem. The present-day version of Bayes' Theorem is the result of further development by the mathematician Pierre-Simon Laplace, who first published the modern formulation in his *Théorie analytique des probabilités*.

Bayes' Theorem is used to compute the probability of an event from the probabilities of other events that influence it. In particular, Bayes' Theorem can be used to predict the probability of the value of a label of a particular instance with mathematical certainty if one knows the values of the attributes that define the instance and the values of several other probabilities. Given two events (A and B) that affect each other, the statement of Bayes' Theorem is deceptively simple:

$$P(A|B)=\frac{P(A)\times P(B\,|\,A)}{P(B)}.$$

This formula gives the probability that A happens given that B has happened, $P(A\,|\,B)$. That probability—the probability of occurrence of an event A, conditioned on the occurrence of event B—plays an important role in Bayesian statistics. I will illustrate its role with the help of the data in table 5.1.

Let us define event A as "a day good for playing tennis" and event B as "a non-windy day, with wind below 15 knots." If we assume that the data are representative of the distribution of day characteristics and that they faithfully reproduce the probabilities of the events, we can conclude that $P(A) = 3/7$, because out of the seven days registered only three were good days on which to play tennis. However, if we consider only non-windy days, we obtain a value of 3/5 for $P(A \mid B)$. This is the probability that a day is good for playing tennis, conditioned on the fact that the day isn't windy.

Before we consider applying this formula to the inductive learning problem, it is instructive to consider a simpler example in which the application of Bayes' formula provides the right way to reason about probabilities.

Suppose that someone randomly selected from the population of the United States decides, for no particular reason, to take a AIDS test. He or she identifies a reputable clinic providing a test that is 98 percent precise. This means the test will give a wrong result only 2 percent of the time, returning either a false positive or a false negative. We know that the incidence of AIDS in the US is estimated to be 0.4 percent, meaning that about four people in a random sample of 1,000 will have AIDS.

Now imagine that the test result is positive. What is the probability that the person has AIDS? One would think it is pretty high, given that the test very seldom gives wrong results. The exact probability that the person has AIDS is given by Bayes' Theorem, with A meaning "has AIDS" and B meaning "test result is positive for AIDS." What, then, are the values of $P(A|B)$ and $P(\bar{A} \mid B)$, which are respectively the probability of the person's having AIDS and the probability of the person's not having AIDS, given that the test result was positive?

As has been noted, the probability of AIDS in the general population, $P(A)$, is 0.004. Therefore, $P(\bar{A}) = 0.996$. From the probability of error of the test results, we know that $P(B|A) = 0.98$, and $P(B|\bar{A}) = 0.02$. These values are, respectively, the probability that the test returns a true positive result (i.e., the person has AIDS and the test result is positive) and the probability that the rest returns a false positive result (i.e., the person does not have AIDS and the test result is positive).

Now, by computing $P(A|B) = P(A) \times P(B \mid A)$ and $P(\bar{A}|B) = P(\bar{A}) \times P(B \mid \bar{A})$, we obtain, respectively, 0.0039 and 0.020. The normalizing factor in the denominator of Bayes' formula, $P(B)$, is the same in both cases and serves only to make these two values add up to 1, since the person either does or doesn't have AIDS. By normalizing the values 0.0039 and 0.020 so that they add up to 1, we obtain the counterintuitive result that, even after testing positive for AIDS, the person still has only a 16 percent probability of having AIDS, and an 84 percent probability of not having it.

Bayes' Theorem enables us to combine the *a priori* probabilities of events with the additional evidence (in this case, the test) that changes our degree of belief in the occurrence of these events, giving us the *a posteriori* probability. For instance, earthquakes are rare, and therefore you will think first of other explanations if you feel the floor of the building you are in shaking. However, if you see a whole building shaking, the added evidence overrides the low probability of earthquakes and you probably will believe that one is happening.

Let us apply Bayes' formula to the example in table 5.1. Consider that event A represents "a day good for playing tennis" and its opposite, $\bar{A}$, represents "not a good day on which to play tennis." We will consider a simplified version of the problem and will use only the first column in table 5.1 to define event B. Event B, therefore, means a warm day (which we will define for this purpose as a day with temperature above 50°F), and event $\bar{B}$ means a cold day (below 50°F). In this simplified version of table 5.1, we can now compute all the values required to apply Bayes' Theorem.

From the table, we can now estimate the following probabilities: $P(A) = 3/7$, because, as we have seen, three days out of seven were good for playing tennis. $P(B|A) = 1.0$, because on every single day that was good for playing tennis it was warm. Finally, $P(B) = 5/7$, because on five days out of seven it was warm. Using these probabilities, we can now apply Bayes' Theorem to compute the probability that a warm day is a good day on which to play tennis. To do so, we compute

$$\frac{(3/7)\times 1.0}{5/7}$$

and obtain 0.6 as the probability $P(A \mid B)$ that a warm day is good for playing tennis. You can verify this value by noticing there are five warm days in figure 5.1, of which three are good for playing tennis.

Note that this computation requires no specific assumption about what makes a day good for playing tennis. We simply computed probabilities from the data and applied Bayes' Theorem to obtain a result, known to be exact. However, there are significant difficulties with the application of

Bayes' Theorem in real life that make it difficult to use directly. These difficulties are related to the computation of the probabilities needed to obtain the value on the right-hand side of Bayes' formula.

Take, for instance, the value of $P(B)$, the probability that a given day is a warm day. Our estimate of this value, 5/7, was obtained from the table, but we have no warranty that it is an accurate estimation of the true value of $P(B)$. We can improve this estimate by obtaining more data or by consulting meteorological records, which would give more accurate estimates of the true value of $P(B)$.

However, if we want to consider all the columns in table 5.1, an even bigger difficulty arises. Suppose we want to compute the probability of an event A, given a conjunction of other events. For instance, we may want to compute $P(A|B \wedge C \wedge D)$, where B is a warm day, C is a dry day, and D is a windy day. Recall that $\wedge$ means "and" and specifies the simultaneous occurrences of these values of the attributes. Now, to apply Bayes' Theorem, we would have to compute $P(A)$, as before, but also $P(B \wedge C \wedge D|A)$. Computing this last probability is difficult and makes it hard to apply. $P(B \wedge C \wedge D|A)$ is the probability that a day good for playing tennis exhibited those specific characteristics, (warm, dry, windy). Computing this probability with some accuracy requires an extensive record and in many cases is not even possible. Even the meteorological records would not be of help, unless they also included the information whether the days were or weren't good for playing tennis. If, instead of three simple attributes, one is dealing with hundreds, such information is never available, because such a particular combination of attributes has probably never occurred before.

Therefore, direct application of Bayes' Theorem to compute the "true" probability of an event cannot be performed, in general. However, the computation can be approximated in many ways, and this leads to many practical classifiers and learning methods. One simple such method, called the *Naive Bayes Classifier*, assumes that, instead of the difficult-to-compute probability $P(B \wedge C \wedge D|A)$, one can use a "naive" approximation that, for a given class, assumes that the values of the attributes are independent. This means that $P(B \wedge C \wedge D|A)$ is replaced by $P(B \mid A) \times P(C \mid A) \times P(D \mid A)$, which is easy to compute because each of the factors can be easily estimated from the table of instances or from other records. In our case, this assumption is equivalent to the statement that, on days that are good for playing tennis, the probability of having a warm, dry, and windy day can be obtained by multiplying the probability of a day being warm by the probability of the day's being dry by the probability of a day's being windy. To be honest, it is usually not the case that the correct result is obtained by multiplying these probabilities. For simplicity, imagine that half of the days good for playing tennis are warm, half of them are dry, and half of them are windy. This assumption then implies that exactly one eighth of days good for playing tennis are warm, dry and windy, because $P(B \wedge C \wedge D \mid A)$ would be approximated by $P(B \mid A) \times P(C \mid A) \times P(D \mid A)$. This assumption probably isn't true, because temperature, dryness, and windiness are correlated and the probability that a day is warm, dry, and windy cannot be approximated by simply multiplying the three separate probabilities.

Despite the fact that the simplifying assumption used by the Naive Bayes Classifier is not true in many cases, this classifier is powerful, is useful in many problems, and, in many cases, yields results that are surprisingly good in comparison with those obtained by more sophisticated methods (Domingos and Pazzani 1997).

Brains, Statistics, and Learning

Using statistics in machine learning to infer the appropriate actions when faced with unseen situations may seem somewhat contrived, artificial, and quite unrelated to what biological systems do. Surprisingly, this is not the case.

Imagine a table somewhat similar to table 5.1, but with the weather attributes replaced by other attributes and the values of the label by other actions. Such a table might represent some more complex problem (perhaps one crucial for survival), and might resemble table 5.2. In such a case, the choice of an action will affect the very survival of the individual.

Table 5.2
Hypothetical decision problem for early humans.

Lion	Rabbit	Leopard	Elephant	Tree	Action
Visible	Near	Visible	Not seen	Near	Run
Visible	Far	Not seen	Not seen	Far	Run
Not seen	Near	Not seen	Visible	Far	Chase
Not seen	Near	Not seen	Not seen	Near	Chase

It is now believed that the brain is, in fact, a very sophisticated statistical machine that keeps an internal model of the relevant parts of the world, and uses it to estimate the probabilities involved in Bayes' Theorem and take the appropriate actions.

The probabilistic models used by the brain are much more complex than those discussed above, and they are not restricted to the evaluation of instances defined by static attributes. Machine learning techniques that take time into account and create internal models of the results of sequences of actions have been used extensively in many machine learning applications, in robotics, in game playing, and in optimization. Reinforcement learning (Sutton and Barto 1998)—in particular, a form of reinforcement learning called *temporal difference learning* (Sutton 1988)—is one such technique. Reinforcement learning is based on the idea that each particular action has an expected reward. Such a reward can be estimated from many action-reward pairs, even if the rewards come with significant delays relative to the actions. It is highly likely that estimation processes that obtain similar results take place inside every living brain, creating the ability to estimate the result of every action and using it to choose the best actions.

The concept that the brain is essentially a very complex statistical machine may be due to Kenneth Craik, who suggested in his 1967 book *The Nature of Explanation* that organisms carry in their brains a "small-scale model" of reality that enables them to capture statistical relationships between events in the outside world. Understanding how brains process sensory information to construct their internal models of the world has been one of the central goals of neuroscience—a goal that remains to be accomplished after many decades of research.

With its ability to compute complex statistical information, the human brain, the most powerful of all known brains, is probably the most sophisticated design crafted by a process that has been going on for billions of years. That process is, in a way, the most complex and longest-running algorithm ever executed. Discovered by Charles Darwin, it is called *evolution*.

6 Cells, Bodies, and Brains

The work that led Charles Darwin to write *On the Origin of Species by Means of Natural Selection* probably didn't begin as an effort to start what would become the major scientific revolution of all time, a paradigm shift that would change people's views of the world more deeply than any other in human history.

It is not likely any other idea in human history has had such a large impact on the way we think today. Many other important paradigm shifts have changed people's view of the world profoundly, among them Copernicus' removal of the Earth from the center of the universe, Newton's laws of motion, Einstein's theory of relativity, the atomic theory, quantum mechanics, the internal combustion engine, and integrated circuits. However, none of those breakthroughs changed the way we view the world and ourselves as profoundly as the theory that Darwin developed in order to explain why there are so many different species on our planet.

The Ultimate Algorithm

Darwin's early interest in nature led him to pursue a passion for natural science, which culminated with his five-year voyage on the *Beagle*. The publication of his journal of the voyage made him famous as a popular author. Puzzled by the geographical distribution of wildlife and the types of fossils collected on the voyage, Darwin conceived his theory of natural selection. Other scientists, among them Alfred Wallace, had arrived at similar conclusions, but Darwin was the first to publish them. In fact, when Darwin became aware of Wallace's ideas, he had to rush to complete and publish his own work, which he had been developing at a more leisurely pace until then.

The simple but powerful idea that Darwin described is that different species of living beings appear because of selective pressure acting differentially on the reproductive success of individuals by selecting qualities that are advantageous, even if only to a small degree:

> If then we have under nature variability and a powerful agent always ready to act and select, why should we doubt that variations in any way useful to beings, under their excessively complex relations of life, would be preserved, accumulated, and inherited? Why, if man can by patience select variations most useful to himself, should nature fail in selecting variations useful, under changing conditions of life, to her living products? What limit can be put to this power, acting during long ages and rigidly scrutinising the whole constitution, structure, and habits of each creature,—favouring the good and rejecting the bad? I can see no limit to this power, in slowly and beautifully adapting each form to the most complex relations of life. The theory of natural selection, even if we looked no further than this, seems to me to be in itself probable. (Darwin 1859)

The central point of Darwin's argument is that species tend to reproduce themselves in accordance with an exponential law because, in each generation, an individual can give rise to multiple descendants. Darwin understood that a single species, growing at an exponential rate, would fill the planet and exhaust all available resources if natural selection didn't limit the growth of populations. Competition between species ensures that those better adapted to the environment have a competitive advantage, measured by the number of surviving offspring. Differences (perhaps small ones) in the rate of growth of populations with different characteristics led, over time, to the selection of advantageous traits. The same principle could be applied as naturally to the feathers of peacocks, the tails of dolphins, the necks of giraffes, and the human brain.

More damaging to the prevailing view that the human species had a central role on Earth, if not in the universe, was the fact that this argument could be applied as well to the appearance of all the characteristics we recognize in humans. In *On the Origin of Species*, Darwin didn't even make this view explicit. He simply alluded, en passant, that "light will be thrown on the origin of man and his history," thus recognizing that the human species was no different from the other species, since it also evolved as a result of the pressure imposed by natural selection.

That Darwin formulated the concept of evolution by natural selection in such a clear way that his ideas remain unchanged by 150 years of scientific development is, in itself, remarkable. We now understand that evolution by natural selection, as proposed by Darwin, can occur only if discrete units of genetic inheritance are somehow able to replicate themselves in the bodies of the offspring. These units, which we now call *genes*, are passed from generation to generation almost unchanged.

Even more remarkable is that Darwin reached his conclusions, which are still valid today, without having the slightest hint about the underlying mechanisms that enabled specific characteristics to be passed down from parents to their descendants. Because he was unaware of contemporary work by Gregor Mendel (see Mendel 1866), Darwin had few ideas—and the ones he had weren't correct (Charlesworth and Charlesworth 2009)—about the physical mechanisms that were behind the working of genes, the discrete units of inheritance necessary for the theory of evolution to work. If physical characteristics were passed from parents to descendants by means of continuous analog encoding, the process of evolution probably wouldn't have started in the first place.

In most species, individuals reproduce by mating, generation after generation. For that reason, even if a specific characteristic such as height were to be selectively advantageous, it would become diluted over successive generations if the genetic makeup of individuals weren't encoded in a discrete way. To understand why this is so, imagine a set of glasses full of dyed water. Some glasses have more dye, some have less. Now imagine that selection acts upon this set of glasses by selecting, for instance, those containing darker-colored water. To obtain the next generation, the glasses with the darker water are selected and combined among themselves. The next generation will, indeed, be more heavily dyed than the original, and the population of glasses, as a whole, will contain darker water than the original population. However, since the water in the glasses was combined to obtain the next generation, this generation is much more uniform than the preceding one. As generations go by, all the glasses will become the same color, and no further darkening of the water can take place.

The story is very different if the amount of dye in each glass is recorded with a discrete encoding, and it is this discrete number that is probabilistically passed to the descendants, not the dyed water itself. Now the problem of convergence of the population to the average of the best individuals is avoided, since different concentrations of dye, even the highest, can still be present in the population many generations from the initial one. Add to this a somewhat rare, but not impossible, event, such as a mutation, and the conditions for endless and open-ended evolution are in place.

The actual mechanisms that define genetic inheritance were brought to light through the work of Mendel. Mendel, the father of genetics, became an Augustinian friar in order to obtain an education. He conducted his studies on the variation of plants by cultivating, between 1856 and 1863, almost 29,000 pea plants. He selected and studied seven plant characteristics, including seed form, pod form, seed color, pod color, and flower color. We now know that it is the encoding of these traits by genes in different chromosomes that leads to the independence between traits that characterizes Mendel's model.

Mendel's extensive experimental results led him to identify the mechanisms that are now known as *Mendel's laws of inheritance*. His results, published in an obscure and largely unnoticed journal (Mendel 1866), showed that genetic traits are passed to the descendants in a digital, on/off way, and also that the organisms he worked with (pea plants) have two copies of each gene, one of which is passed to the following generation. In fact, Mendel's results apply to all diploid organisms (i.e., organisms with two copies of each chromosome). Most animals and plants are diploid. Some, however, are polyploid—a salmon, for example, has three copies of each chromosome.

Mendel further deduced that different traits are independently passed to the next generation—something that was true of the traits he studied, but that, in general, applies only to characteristics encoded by independent genes. It is worthwhile to consider his results in more detail.

Suppose that a certain trait in a plant is encoded by a gene that can take one of only two values, called *alleles*: the value *A* (which we will assume dominant) or the value *a* (which we will assume recessive). A specific value of an allele is called *dominant* if it masks the other value the allele can take. In this example, we are assuming that there are only two alleles for each gene and two copies of each chromosome, and thus we have four possible values in each locus: *AA*, *Aa*, *aA*, and *aa*. Since allele *A* is dominant, the plants with the *AA*, *Aa*, and *aA* variants will exhibit the trait, and the plants with the *aa* variant will not. When a plant reproduces, the offspring inherit one copy of the gene from each of the progenitors. With this information at hand, we can proceed with the analysis performed by Mendel, which remains the basis of modern genetics.

We begin the experiment with some plants with the *AA* variant and some with the *aa* variant. When we cross an *AA* plant with an *aa* plant, we obtain plants with *Aa*, which, like the *AA* parent, exhibit the trait because the *A* allele dominates the *a* allele. Things get more interesting in the second generation, when we cross the *Aa* plants with other *Aa* plants. Since the next generation will inherit one gene from each parent, we will now get *AA* plants (roughly 25 percent), *Aa* and *aA* plants (roughly 50 percent), and *aa* plants (roughly 25 percent). Since the *AA* and *Aa* plants exhibit the trait whereas the *aa* plants do not, we expect a ratio of 3:1 between the plants that exhibit the trait and the plants that do not. This is exactly the result that Mendel obtained.

Mendel's careful experimental procedure led to a very conclusive verification of the hypothesis that traits are passed on to the next generation in a discrete way by combining the genes of the parents. In fact, the conclusions he reached were so clear, precise, and convincing that, many years later, a dispute arose about the plausibility of the extremely high precision of his experimental results. The statistician and geneticist Ronald Fisher (1936) pointed out the experimental ratios Mendel obtained in his experiments were implausibly close to the ideal ratio of 3:1. Fisher analyzed the results that led to the 3:1 ratio in the second generation and found that rerunning the experiments would give results closer to this theoretical ratio less than once per thousand attempts. Thus, either Mendel was extremely lucky or there was something to be explained about Mendel's results. The controversy is far from over, as is witnessed by the many books and articles on the subject still being published. Although few people would accuse Mendel of doctoring the data, the results have continued to be mysterious for many. Among many explanations, it has been suggested Mendel may have either dropped or repeated some results that were not considered sufficiently consistent with his hypothesis (Franklin et al. 2008; Pires and Branco 2010).

Mendel's work was largely ignored in his time, and it wasn't widely accepted until many years after his death, when the need for a discrete model for inheritance became clear. The physical and molecular reasons for Mendel's results would not become clear until 90 years later, when James Watson and Francis Crick (1953) identified deoxyribonucleic acid (DNA) as the molecule responsible for genetic inheritance and discovered its helical structure.

Despite the minor difficulty caused by the fact that no mechanism, known at the time, could support the type of inheritance necessary for evolution to work, Darwin's ideas survived all discussions that ensued. In present-day terms, Darwin had simply proposed that life, in all its diversity, was the result of a blind algorithmic process that had been running on Earth for about 4 billion years. That evolution can be seen as an algorithm should come as no surprise to readers now familiar with the concept. Daniel Dennett (1995) expressed this idea clearly and beautifully:

> The theoretical power of Darwin's abstract scheme was due to several features that Darwin firmly identified, and appreciated better than many of his supporters, but lacked the terminology to describe explicitly. Today we would capture these features under a single term. Darwin had discovered the power of an algorithm.

The idea that life on Earth, in all its forms, was created by an algorithm—a process that blindly, step by step, applied fixed and immutable rules to derive the immense complexity of today's world—caught people by surprise. The ages-old question of how the different species appeared on the planet had, after all, a simple and self-evident answer. Natural selection sufficed to explain the wide variety of species found in the tree of life. To be fair, many questions remained unanswered. The most vexing one was, and still is, related to the question of how evolution began. After all, natural selection can be used to explain differential rates of reproduction only after the reproduction scheme is in place, after organisms that reproduce and exhibit variations exist. Darwin didn't try to answer that question. Even today, we have only the foggiest idea of what were the first systems that could, in some way, replicate themselves.

Many arguments have been made against the theory of evolution, and there are still many skeptics. Creationists believe that life was created, in its present form, by a divine entity, perhaps only a few thousand years ago. Of the many arguments that have been leveled at the theory of evolution, the strongest is the argument that such complex designs as we see today could not have originated in a succession of random changes, no matter how sophisticated the blind algorithm conducting the process. At some point in the course of billions of years, the argument goes, some intelligent designer intervened. If we see a watch (a complex device), we naturally deduce that there is a watchmaker. Yet, the argument continues, we look at an eye—something far more complex than a watch—and believe it has evolved by an almost random accumulation of small changes. Many authors have presented beautiful answers to this argument, and I could do no better than quote one of the most eloquent of them, Richard Dawkins (1986):

> Natural selection is the blind watchmaker, blind because it does not see ahead, does not plan consequences, has no purpose in view. Yet the living results of natural selection overwhelmingly impress us with the appearance of design as if by a master watchmaker, impress us with the illusion of design and planning.

And yet, even though most of us are ready to accept that evolution, a blind process, has created all creatures, big and small, we may still be somewhat convinced that the end product of this process is the creature known as *Homo sapiens*. This is very unlikely to be the case. Evolution did not set out to create anything in particular. If the process was to be run all over again, it is extremely unlikely that something like the human species would come out of it. Too many accidents, too many extremely unlikely events, and too much randomness were involved in the evolution of complex creatures. The development of the first self-reproducible entity, the evolution of eukaryotic cells, the Cambrian explosion, the extinction of dinosaurs, and even the fact that *Homo sapiens* was lucky enough to survive many difficult challenges and evolutionary bottlenecks (unlike its cousins) are just a few examples of historical events that might have ended differently. Had they not occurred, or had they ended in a different way, the history of life on Earth would have been very different, and Earth might now be populated by an entirely different set of species. In the words of one famous evolutionary biologist,

> We are glorious accidents of an unpredictable process with no drive to complexity, not the expected results of evolutionary principles that yearn to produce a creature capable of understanding the mode of its own necessary construction. (Gould 1996)

By the same reasoning, it is extremely unlikely that *Homo sapiens* is the final result of this multi-billion-year process. However, it is true that, for the first time in the history of the planet, one species has the tools to control evolution. We have been controlling evolution for centuries, by performing human-driven selection of crops and of dogs, horses, and other domestic animals. But today we are on the verge of developing technologies that can be used to directly control the reproductive success of almost all forms of life on Earth, and we may even become able to create new forms of life.

It is important to remember that natural selection doesn't evaluate and reward the fitness of individuals. True, the number of descendants is statistically related to the fitness of a specific individual. But, given the mechanisms of genetic inheritance, and the dimension of the Library of Mendel (which was discussed in chapter 2), even very fit individuals are likely to exist only once in the whole history of the universe. No future copies of exceptional individuals will be selected for future reincarnation by natural selection. In general, at least in organisms that reproduce sexually, the offspring will be very different from even the most successful individuals. Instead, specific versions of those elusive entities called genes will be preserved for future generations. Though it may be intuitive to think that natural selection is selecting individuals, it is more reasonable to view the gene as the basic unit of selection.

The genes were the original entities that, roughly 4 billion years ago, managed to find some process that would replicate them. They became, therefore, replicators. Successful genes became more common, and, therefore, increased in number at an exponential rate, if selective pressure didn't act to curtail their expansion. The result is that the most common genes in a population are those that, in the average, through a large number of genotypes and in a large number of situations, have had the most favorable effects on their hosts. In other words, we should expect selfish genes to flourish, meaning that they will promote their own survival without necessarily promoting the survival of an organism, a group, or even a species. This means that evolutionary adaptations are the effects genes have on their hosts, in their quest to maximize their representation in the future generations. An adaptation is selected if it promotes host survival directly or if it promotes some other goal that ultimately contributes to successful reproduction of the host.

With time, these genes encoded more and more complex structures—first simple cells, then complex eukaryotic cells, then bodies, and finally brains—in order to survive. In fact, organisms, as we know them today, exist mainly to ensure the effective replication of these units, the genes (Dawkins 1986). In the most extreme view, organisms exist only for the benefit of DNA, making sure that DNA molecules are passed on to the next generation. The lifetimes of genes—sequences of DNA encoding for a specific protein or controlling a specific process—are measured in millions of years. A gene lasts for many thousands of generations, virtually unchanged. In that sense, individual organisms are little more than temporary vehicles for DNA messages passed down over the eons.

We may think that, with the passage of time, things have changed and evolution is now centered on organisms, and not on the genetic units of inheritance we now know to be encoded in the form of specific combinations of DNA bases. We may think these original and primitive replicators are now gone, replaced by sophisticated organisms, but that would be naive. Perhaps many millions of years after mankind is gone from the Earth, the most effective replicators will still populate the planet, or even the known universe. We have no way of knowing how these replicators will look millions of years from now. Maybe they will be very similar to today's genes, still encoding specific characteristics of cells—cells that will form organisms much like the ones that exist today. But I do not believe that. One of my favorite passages by Dawkins (1976) makes it clear we are no more than vehicles for the survival of genes, and that we should be aware that they are ingenious entities indeed:

> Was there to be any end to the gradual improvement in the techniques and artifices used by the replicators to ensure their own continuation in the world? There would be plenty of time for their improvement. What weird engines of self-preservation would the millennia bring forth? Four thousand million years on, what was to be the fate of the ancient replicators? They did not die out, for they are the past masters of the survival arts. But do not look for them floating loose in the sea; they gave up that cavalier freedom long ago. Now they swarm in huge colonies, safe inside gigantic lumbering robots, sealed off from the outside world, communicating with it by tortuous indirect routes, manipulating it by remote control. They are in you and me; they created us, body and mind; and their preservation is the ultimate rational for our existence. They have come a long way, those replicators. Now they go by the name of genes, and we are their survival machines.

Given the characteristics of exponential growth of entities that reproduce themselves at some fixed rate, we can be sure that replicators, in some form, will still be around millions of years from now. Many of them may no longer use DNA as their substrate to encode information, but some of them will certainly remain faithful to that reliable method.

The essence of a replicator is not the medium used to store the information; it is the information that is stored. As long as mechanisms to replicate the entities are available, any substrate will do. In the beginning, DNA probably wasn't the substrate of choice, because it requires complicated reproduction machinery that wasn't available then. In fact, there is another molecule that is extensively used to store and transmit genetic information inside the cells: ribonucleic acid (RNA). It may have been a precursor of DNA as the substrate to store biological information. RNA plays various biological roles inside cells. Unlike DNA, it is usually a single-stranded coil.

According to the RNA World Hypothesis, RNA molecules were precursors of all current life on Earth before DNA-based life developed. That widely discussed hypothesis is favored by many researchers because there is some evidence that many old and very stable cellular structures are based on RNA. However, even supporters of the RNA World Hypothesis believe that other, even less developed supports probably existed before RNA. What they were remains, however, an open question.

In the future, other substrates may be better able than DNA to store and process the information required by replicators to survive and reproduce. We are all familiar with computer viruses, and know well the potential they have to replicate themselves, in an exponential fashion, when the right conditions are present. And yet, these computer viruses are very primitive replicators, brittle and unsophisticated. In the future, much more sophisticated entities that can replicate themselves in the memories and CPUs of computers may come into existence. These may be the replicators of the future, and they will be as similar to today's DNA-based creatures as today's creatures are similar to the replicators that populated the warm ponds on Earth 4 billion years ago.

Cells and Genomes

Now that we understand the algorithm that led to evolution by natural selection and are aware of the existence of those somewhat mysterious units of inheritance called genes, we can proceed to understand how the basic replicators of yore, the genes, managed to evolve the complex structures that are now used to keep them alive: multi-cellular organisms.

Even after the publication and assimilation of Darwin's work, biology remained a relatively calm and unfashionable field of research. The principle of evolution by natural selection and the basic tenets of genetics opened the door to the treatment of biology as a science based on first principles. For a long time, however, ignorance of the physical mechanisms that supported life and ignorance of genetics stood in the way of a truly principled approach to biology. In fact, for most of the twentieth century the fundamental mechanisms underlying evolution and speciation were unknown, since the biochemical basis for life was poorly understood, and little was known about the way genetic characteristics were passed from generation to generation. In fact, until 1953 no one knew exactly how genetic characteristics were passed from parents to their descendants.

While Darwin worked on the theory of evolution, a number of biologists were homing in on the central importance of the cell in biology. In 1838 Theodor Schwann and Matthias Schleiden began promoting the idea that organisms were made of cells and that individual cells were alive. (See Schwann 1839; Schleiden 1838.) By about 1850, most biologists had come to accept that cells were the basic units of living beings, were themselves alive, and could reproduce. By 1900, many pathways of metabolism, the way molecules interact with each other, were known. Improved techniques such as chromatography (mixture-separation techniques based on the use of gels) and electrophoresis (the use of electromagnetic fields to separate differently charged particles) led to advances in biochemistry and to a reasonable understanding of some of the basic biochemical mechanisms found in cells.

Oswald Avery, Colin MacLeod, and Maclyn McCarty (1944) demonstrated that DNA was the substance responsible for producing inheritable change in some disease-causing bacteria and suggested that DNA was responsible for transferring genetic information to offspring. However, not until 1953, when Watson and Crick demonstrated that the DNA molecule was the carrier of genetic information, did genetics begin to play the central role it has today. Until then, the role of DNA was mostly unknown. The biophysicist Max Delbrück even called it a "stupid molecule," noting that it monotonously repeated the four bases A, C, T, and G (for adenine, cytosine, thymine, and guanine), whereas proteins were known to be extraordinarily versatile in form and function. Watson and Crick, using x-ray data collected by Rosalind Franklin and Maurice Wilkins, proposed the double-helix structure of the DNA molecule, thereby solving a centuries-old question about how genetic information was passed from cell to cell and from parents to their descendants.

When Watson and Crick identified the double helix of DNA as the repository for genetic information, stored in digital format in a molecule that until then hadn't been thought of as very exciting, they opened the door to the most exciting decades in the history of biological sciences. Suddenly, the discrete encoding mechanism of genetics inherent to Mendel's ideas and to Darwin's evolution mechanism was discovered, in the form of a molecule with a simple, repetitive structure. Each segment of each one the strands of the double helix of a DNA molecule contains one base, or nucleotide, encoding one of possible four values, A, C, T, and G. The nature of the chemical bonds between the two strands of the helix makes sure that an A is always paired with a T, and that a C is always paired with a G, as shown in figure 6.1. This redundancy makes it possible to create two copies of a DNA molecule by splitting the double helix into two single helices, which then act as templates for two new, double-stranded DNA molecules.

This giant step in our understanding of the way nature passed characteristics from a cell to other cells, and from an organism to its descendants, eventually led to another revolution—a revolution that is still underway. With this discovery, it became finally clear that the replicators in today's world are made of DNA, and that this molecule is at the center of all cellular activity. It is DNA that carries the genetic information as cells divide, and from parents to offspring. Therefore, DNA must contain, encoded in its

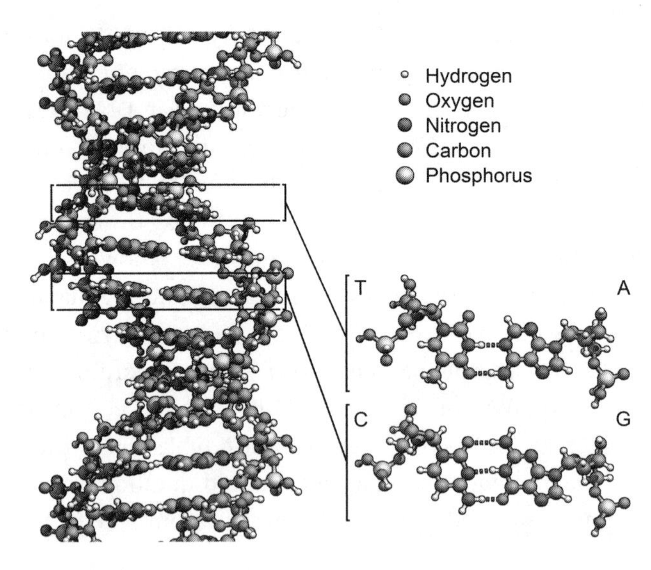

Figure 6.1
The structure of the double helix of the DNA molecule, in which A-T and C-G base pairing creates the redundancy needed for DNA duplication when cells split or organisms reproduce. Drawing by Richard Wheeler (2011), reproduced with permission; also available at Wikimedia Commons.

long sequence of bases, the structures that represent the elementary units of inheritance.

We now know that DNA encodes information used in the cells in many different ways, many of them only partially understood. However, its most central role is the encoding of the sequence of amino acids in proteins. Proteins are sequences of amino acids that, once formed, fold in complex three-dimensional structures to perform many different functions inside the cells. They can be thought of as very versatile nano-machines.

The knowledge that DNA encodes the proteins used to build cells, and therefore to build all living beings, opened the doors to enormous advances in our understanding of cells and biological systems. Since DNA (which stores, in each position, only one of four possible values) is used to encode the sequence of each protein, and there are twenty amino acids used by proteins, it was soon postulated that sets of three bases (at least) would be required to encode each amino acid. In fact, two nucleotides could encode only sixteen (4^2) different combinations and could not be used to encode each of the twenty amino acids used in proteins.

Additional experiments by a number of scientists led to the discovery of the genetic code, which specifies how each set of three DNA bases encodes a specific amino acid. Marshall Nirenberg and Philip Leder (1964), building on work done by Crick et al. (1961) which demonstrated that three bases of DNA code for one amino acid, deciphered the codons of the standard genetic code.

In figure 6.2, the entries marked "Start" and "Stop" signal the beginning and the end of each protein sequence, although we now know that, in reality, the process of translation from DNA to protein is significantly more complex, and other mechanisms may be used to initiate or terminate translation.

We now know that some organisms use slightly different genetic codes than the one shown in figure 6.2. Barrell, Bankier, and Drouin (1979) discovered that the constituents of human cells known as mitochondria use another genetic code. Other slight variants have been discovered since then, but the standard genetic code shown in figure 6.2 remains the reference, used in the majority of the cases.

With the discovery of the genetic code and the basic mechanisms that lead from DNA to protein and to cells, the complete understanding of a biological system becomes possible and, apparently, within reach. With

First letter	Second letter: U	C	A	G	Third letter
U	Phenylalanine	Serine	Tyrosine	Cysteline	U
U	Phenylalanine	Serine	Tyrosine	Cysteline	C
U	Leucine	Serine	Stop	Stop	A
U	Leucine	Serine	Stop	Tryptophan	G
C	Leucine	Proline	Histidine	Arginine	U
C	Leucine	Proline	Histidine	Arginine	C
C	Leucine	Proline	Glutamine	Arginine	A
C	Leucine	Proline	Glutamine	Arginine	G
A	Isoleucine	Threonine	Asparagine	Serine	U
A	Isoleucine	Threonine	Asparagine	Serine	C
A	Isoleucine	Threonine	Lysine	Arginine	A
A	Methionine, Start	Threonine	Lysine	Arginine	G
G	Valine	Alanine	Aspartic acid	Glycine	U
G	Valine	Alanine	Aspartic acid	Glycine	C
G	Valine	Alanine	Glutamic acid	Glycine	A
G	Valine	Alanine	Glutamic acid	Glycine	G

Figure 6.2
The standard genetic code. Each codon is composed of three bases and encodes one amino acid.

time, it was indeed possible to identify structures in DNA, the genes, which encoded specific proteins, the molecular machines central to cellular behavior. However, in contrast with the cases Mendel studied, most genes do not have a direct effect on a visible trait of the organism. Instead, genes encode for a specific protein that will have many functions in the cells of the organism. Still, the understanding of the structure of the DNA, coupled with the development of many techniques dedicated to the study of cells, led to a much better understanding of cells. The quest to understand in a detailed and principled way how cells work was one of the most fascinating endeavors of the last four decades of the twentieth century, and it will certainly be one of the most interesting challenges of the twenty-first.

A cell is a highly complex machine, crafted by evolution over several billion years. All cells share common ancestors: single-cell organisms that were present at the origin of life roughly 4 billion years ago. One of these organisms must be the most recent common ancestor of all existing life forms. It is known by the affectionate name LUCA, standing for Last Universal Common Ancestor. LUCA is believed to have lived more than 3 billion years ago. We know, from the analysis of DNA sequences, that all the organisms in existence today, including bacteria, archaea, and eukarya (figure 6.3), evolved from this common ancestor.

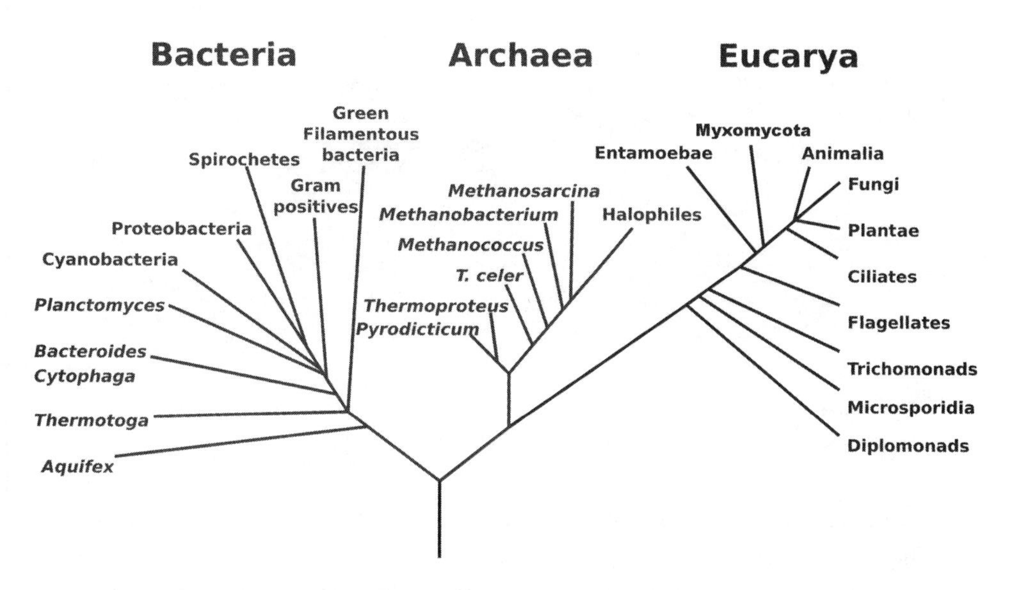

Figure 6.3
A phylogenetic tree of life based on genomic data. Source: Woese 1990 (available at Wikimedia Commons).

Bacteria and archaea are prokaryotes (simple single-celled organisms). They are very numerous and proliferate in almost any imaginable environment. All multi-cellular plants and animals, including humans, belong to the eukarya domain. They are eukaryotes, meaning they are built out of eukaryotic cells. Eukaryotic cells are estimated to have appeared sometime between 1.5 billion and 2.7 billion years ago (Knoll et al. 2006; Brocks et al. 1999), after several billion years of evolution of prokaryotic cells. Eukaryotic cells probably originated in a symbiotic arrangement between simpler prokaryotic cells that, in the end, became a single, more complex cell. We don't know exactly how eukaryotic cells appeared, but they represent a major event in the development of complex life. In a prokaryotic cell the majority of the genetic material is contained in an irregularly shaped region called the *nucleoid*. The nucleoid is not surrounded by a nuclear membrane. The genome of prokaryotic organisms is generally a circular, double-stranded piece of DNA, of which multiple copies may exist.

Eukaryotic cells are more complex and are composed of a large number of internal membranes and structures, called *organelles*. A detailed description of the structure of a eukaryotic cell would require several large

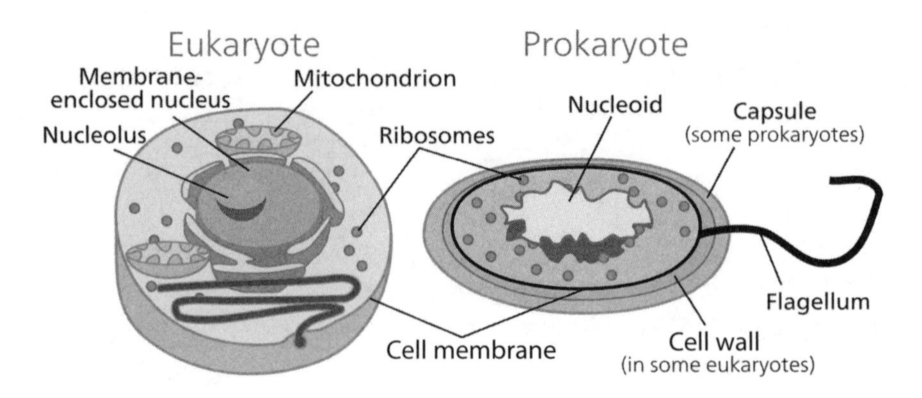

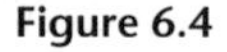
Figure 6.4

Simplified views of eukaryotic and prokaryotic cells (NCBI 2007).

volumes. A number of excellent references exist that describe in more detail how cells work; one is Cooper and Hausman 2000. Here, I will give only a brief and simplified overview of the general architecture of eukaryotic cells, using figure 6.4 as an illustration.

The most central structure in a eukaryotic cell is the nucleus, which contains the chromosomes and is surrounded by a membrane. Outside the nucleus, in the cytoplasm, are a number of structures that play different roles. Mitochondria are organelles involved in a number of tasks. They are mostly known as the cell's power generators, because they generate the majority of the cell's supply of chemical energy. It is believed that mitochondria were once separate organisms that were absorbed or otherwise incorporated in cells. They have their own separate, DNA, which, in humans is passed mostly along the feminine line. Ribosomes are complex molecular machines, with a mass equivalent to about 3 million hydrogen atoms, which synthesize proteins from messenger RNA in a process known as *translation* (described below). Most eukaryotic cells have a cytoskeleton, composed of microtubules and microfilaments, that plays an important role in defining the cell's organization and shape.

Eukaryotic DNA is divided into several linear bundles called *chromosomes*. The process by which DNA is read and by which its instructions are translated into operating cell machinery, the so-called central dogma of biology, is now relatively well understood, although many aspects remain insufficiently clear. What follows is a simplified and somewhat schematic description of the way DNA code is turned into working proteins in a eukaryotic organism.

The process begins with transcription. During transcription, the information contained in a section of DNA in the nucleus is transferred to a newly created piece of messenger RNA (mRNA). A number of proteins are involved in this process, among them RNA polymerase (which reads the information contained in the DNA) and other auxiliary proteins, including the proteins known as *transcription factors*. Transcription factors are important because not all genes present in the DNA are transcribed at equal rates, and some may not be transcribed at all. The rate of transcription is controlled by the presence of the transcription factors, which can accelerate, reduce, or even stop the transcription of specific genes. This is one of the mechanisms that lead to cell differentiation. Cells with exactly the same DNA can behave differently in different tissues or at different instants in time. In eukaryote cells the primary messenger RNA transcript is often changed via splicing, a process in which some blocks of mRNA are cut out and the remaining blocks are spliced together to produce the final mRNA. This processed mRNA finds its way out of the nucleus through pores in the nuclear membrane and is transported into the cytoplasm, where it binds to a ribosome (the cell's protein-making machine).

The ribosome reads the mRNA in groups of three bases, or codons, usually starting with an AUG codon. Each codon is translated into an amino acid, in accordance with the genetic code. Other molecular machines bring transfer RNAs (tRNAs) into the ribosome-mRNA complex, matching the codon in the mRNA to the anti-codon in the tRNA, thereby adding the correct amino acid to the protein. As more and more amino acids are linked into the growing chain, the protein begins to fold into the working three-dimensional conformation. This folding continues until the created protein is released from the ribosome as a folded protein. The folding process itself is quite complex and may require the cooperation of many so-called chaperone proteins.

Proteins perform their biological functions by folding into a specific spatial conformation, or conformations, imposed by the atom interactions. Interactions between atoms result from a number of phenomena that lead to attractive and repulsive forces. Hydrogen atoms, for instance, form H_2 molecules by sharing electrons between them, and are kept together by what is called a covalent bond, as opposed to the non-covalent bonds that exist between atoms that don't share electrons. Non-covalent bonds result from the electrostatic interactions between charged ions or polar molecules. Molecules of water are formed by one atom of oxygen and two atoms of hydrogen. The arrangement of the atoms in these molecules is such that one part of the molecule has a positive electrical charge while the other part has a negative charge. Such molecules, called *polar molecules*, interact with other polar molecules through electromagnetic forces.

The resulting set of forces that exists either between atoms of the protein or between these atoms and molecules in the environment (most commonly water molecules) imposes a structure on the proteins. Protein structure is an important topic of research in biology. Many techniques, including x-ray crystallography and nuclear magnetic resonance spectroscopy, are used to elucidate the structures of proteins. It is conventional to consider four levels of protein structure: primary, secondary, tertiary, and quaternary.

The primary structure of a protein is the linear sequence of amino acids in the amino acid chain. It is determined by the sequence of DNA that encodes the protein. The secondary structure refers to regular sub-structures in parts of the protein, which lead to specific geometrical arrangements of the amino acid chain. There are two main types of secondary structures: alpha helices and beta sheets (Pauling, Corey, and Branson 1951). Tertiary structure refers to the three-dimensional arrangement of protein molecules. Alpha helices and beta sheets are folded, together with the rest of the amino acid chain, into a compact structure that minimizes the overall energy of the molecule, in contact with the environment (usually either water or other proteins) that surrounds it. The folding is driven by a number of different atomic interactions, which include the positioning of hydrophobic residues away from water, but the structure is mostly defined by specific atom to atom interactions, such as non-covalent bonds, hydrogen bonds, and compact packing of side chains. *Quaternary structure* refers to the interaction between different proteins, and consists in determining the ways different proteins interact and dock with each other. Protein complexes range from fairly simple two-protein arrangements (dimers) to complexes with large numbers of sub-units. The ribosome complex illustrated in figure 6.5 is made up of dozens of proteins.

Many computational methods have been developed to determine, in a computer, the secondary, tertiary, and quaternary structures of proteins.

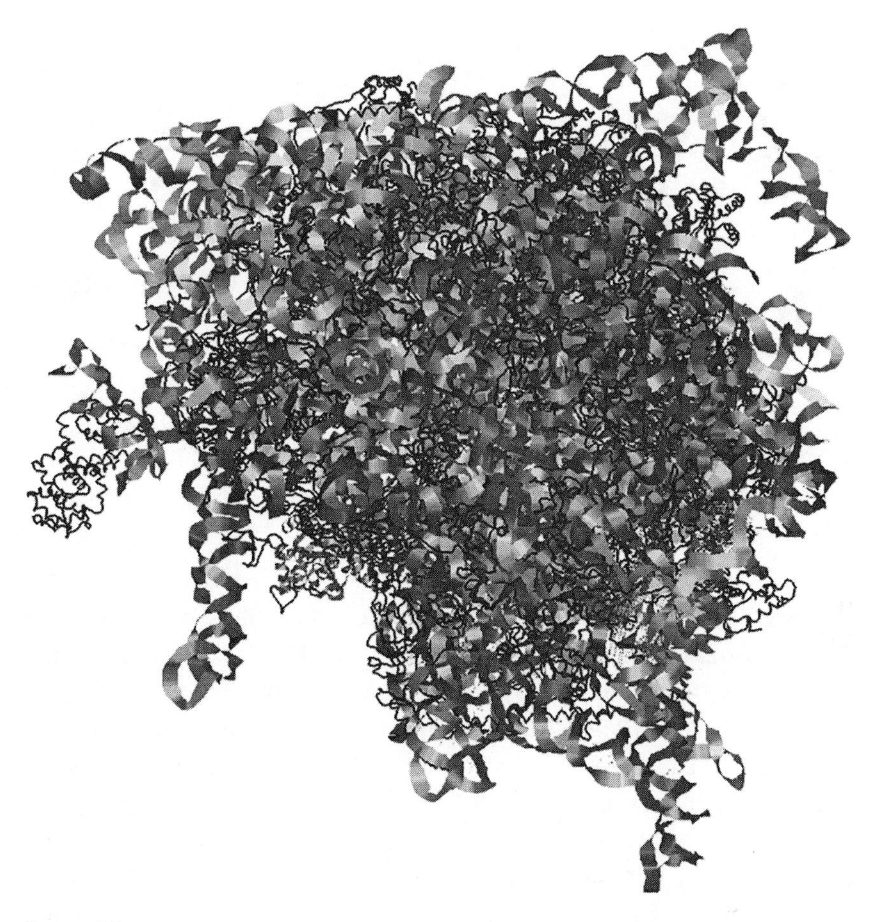

Figure 6.5
The structure of the human 80S ribosome (Anger et al. 2013), from the Protein Data Bank (Berman et al. 2000). Drawn using Jmol (Herraez 2006).

The ability to determine the structures of a given protein is critical for the understanding of many biological phenomena and for the design of new drugs. Today it is possible to predict, in a computer, the spatial configurations of a protein, by considering both its similarity to other proteins and the overall energy of each configuration (Söding 2005; Roy, Kucukural, and Zhang 2010). There are a number of competitions in which different methods are tested and their accuracies are compared. One such competition, known as CASP (Critical Assessment of Structure Prediction), has taken place every two years since 1994.

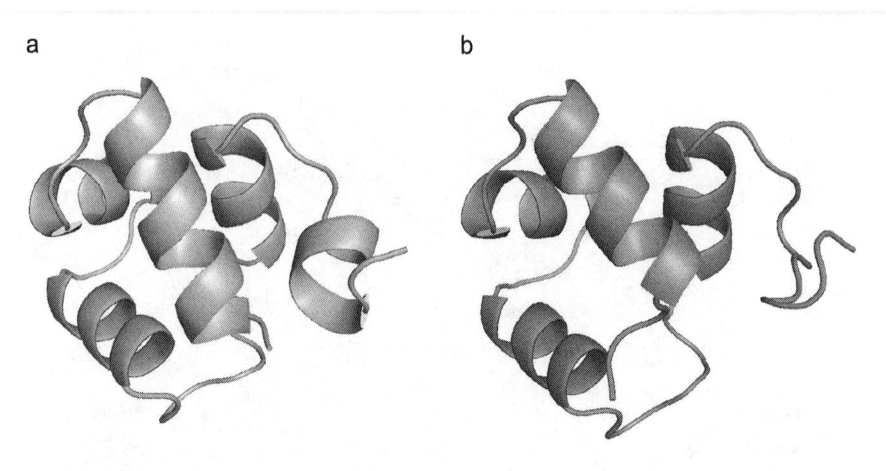

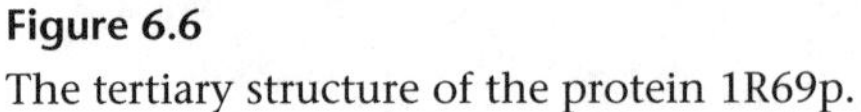

Figure 6.6
The tertiary structure of the protein 1R69p.

However, the problem remains largely unsolved when the protein under analysis is not similar, at least in part, to other known proteins, because the search for the right spatial configuration is, in this case, computationally demanding. Figure 6.6a shows a schematic view of the tertiary structure of the small protein 1R69p, which comprises only five short alpha helices (Mondragon et al. 1989). Even for such a small protein, it isn't easy to determine its spatial configuration *ab initio* (that is, using only the primary sequence and not using information about its similarity with other known proteins). Figure 6.6b shows the result of an *ab initio* determination of tertiary structure of this protein by computational methods (Bugalho and Oliveira 2008) that obtains only an approximate result.

Many other processes, other than transcription and translation, go on inside a cell at any given moment. Some of these processes are reasonably well understood; others are still mostly unknown. Different biochemical processes regulate the flow of nutrients and wastes that enter and leave the cell, control the reproduction cycle, and make sure that adequate amounts of energy are available to enable the cell to operate. Because many processes inside a cell require energy, a significant fraction of the cell's activities are dedicated to obtaining energy. A cell uses the chemical energy stored in adenosine 5´-triphosphate (ATP), generated mostly by the mitochondria, to drive many energetically unfavorable chemical reactions that must take place. In the glycolysis process, which is common to almost all cells, conversion of glucose into pyruvate results in a net gain of two molecules of ATP. This is why glucose, a type of sugar, is needed to drive biological processes in almost all living cells. This process and many other complex metabolic processes are reasonably well understood, and the conversion pathways are well documented (Cooper and Hausman 2000).

Most eukaryotic cells range in size from 2 to 100 μm (micrometers). This may seem small, but we must keep in mind that the radius of an atom of carbon is roughly 70 picometers, and the diameter of a ribosome is on the order of 0.3 nanometers. If a typical eukaryotic cell, with a diameter of 25 μm, were to be blown up to be a mile wide, a carbon atom would be about the size of a blueberry, and a ribosome would be only 6 feet wide. This means that there is enough space in one eukaryotic cell to contain hundreds or even thousands of ribosomes and millions of other cellular machines. Despite the enormous effort invested in understanding cellular behavior, our understanding of cellular mechanisms is still fragmentary and incomplete in all but the simpler processes.

The ultimate objective of biology is to understand not only the behavior of specific cellular processes but also the behavior of organs and even whole organisms. This is done, nowadays, using computers and algorithms to process and model the data obtained by experiments. Before I describe how computers and algorithms can be used to study and model biological systems, I must discuss why bodies and brains became essential for the survival of eukaryotic cells.

Bodies and Brains

Multi-cellular organisms have arisen independently a number of times, the first appearance probably beginning more than a billion years ago (Knoll et al. 2006). A major and sudden increase in the number of the most complex multi-cellular organisms, animals, occurred about 600 million years ago, before and during the Cambrian explosion, a period of dramatic changes in Earth's environment (Grosberg and Strathmann 2007).

Multi-cellular organisms may arise either because cells fail to separate during division or because cells get stuck together accidentally. If natural selection favors this new form of life, it will survive and prosper. There are many possible advantages for multi-cellular organisms. Since there is always an ecological niche for larger organisms, evolution may favor larger species, which are necessarily multi-cellular. As organisms get larger, they have to develop specialized cell types in order to move toward the greater complexity required to sustain larger bodies.

Over time, the blind algorithmic process of evolution led to differentiated cell types and to increasingly complex multi-cellular organisms. Differentiated cells types led to different organs in complex animals, and ultimately to the development of an organ specialized in information processing: the brain. It is now becoming apparent that much of the complexity observed in multi-cellular organisms is due not to major innovations at the level of cellular machinery but, mainly, to gene regulation mechanisms, some of them post-transcriptional, that lead to cell differentiation and specialization. It may be somewhat disheartening to some people to think that their bodies are made of rather common cells—cells very similar to those of many other animals, and even somewhat similar to unicellular organisms such as baker's yeast. In fact, humans have approximately the same number of genes as mice and rats, and have fewer genes than rice and some other species. It isn't likely that there is any specialized set of genes used specifically to make the human brain. Not only do we have approximately the same number of genes as a mouse, but the genes are very similar. Even the humble fruit fly shares many genes with us. The genes that lay out the body plan are very similar in a fruit fly and in a human being, and were inherited from a distant common ancestor—some sort of flatworm that lived about 600 million years ago. The human brain probably is encoded by essentially the same genes that encode monkeys' brains, with some small tweaks in gene regulation that lead to dramatically different brain characteristics.

From the point of view of the replicators, bodies are simply a convenient mechanism to ensure success in replication. Isolated cells, swimming alone in the seas, were too vulnerable to attacks from many sources. The multicellular organism, with highly specialized cells that share the same genetic material, was a convenient way to ensure that many cells were working toward the successful replication of the same genes. Over time, bodies became highly specialized, and brute force was no longer sufficient to ensure survival. The replicators, therefore, came up with a new trick. They mutated in order to encode and generate information-processing cells that could be used to sense the environment and anticipate the results of possible actions. These cells could transmit information at a distance and combine data from different sources to help in decisions. Some of these cells began transmitting information using electrical signals, which travel faster than biochemical signals and can be used in a more flexible way.

Many of the mechanisms needed to transmit electrical signals, and to modulate them through chemical signals, have been found in single-celled organisms known as *choanoflagellates*, which are the closest living relatives of the animals. A choanoflagellate is an organism that belongs to a group of unicellular flagellate eukaryotes with an ovoid or spherical cell body and a single flagellum. The movement of the flagellum propels a free-swimming choanoflagellate through the water.

Choanoflagellates are believed to have separated from the evolutionary line that led to humans about 900 million years ago (Dawkins 2010), and so the mechanisms to transmit electrical signals, if they have the same origin (a likely hypothesis), arose no later than that. Somewhere around that time, cells developed the potential to communicate with other cells by using electrical pulses as well as the chemical signals that were used until then.

With time, neurons—cells able to generate electric impulses—evolved long extensions, which we call *axons*, to carry electrical signals over long distances. Neurons transmit these signals on to other neurons by releasing chemicals. They do so only at *synapses*, where they meet other neurons (or muscle cells). The ability to convey messages using electric signals (instead of only chemical signals) gave the early organisms that possessed these cells a competitive edge. They could react more rapidly to threats and could find food more efficiently. Even a few neurons would have been useful in the fight for survival in the early seas.

Distinguishing light from darkness or having the ability to hear noises is useful; however, being able to sense the outside world, and being able to anticipate the results of specific actions without having to perform those actions, are tricks even more useful for survival. A more sophisticated brain enables an organism to choose the best of several possible actions without running the risk of trying them all, by simply estimating the probability of success of each action. Different connections of neurons, defined by different genetic configurations, evolved to combine information in different ways, some more useful than others, and led some animals to safer or less safe behaviors. The most useful neuron structures proliferated, since they gave their owners a competitive edge, and this information-processing apparatus became progressively more sophisticated as millions of generations went by. What started as simple information-processing equipment that could tell light from darkness, or sound from silence, evolved into complex information-processing circuits that could recognize the presence of enemies or prey at a distance, and could control the behavior of their owner accordingly.

The first neurons were probably dispersed across the bodies of early animals. But as neuron complexes became increasing sophisticated, neurons began to group together, forming the beginnings of a central nervous system. Proximity of neurons meant that more sophisticated information processing could take place rapidly—especially where it was most needed: near the mouth and the light-sensing devices (which would later become eyes). Although it isn't yet clear when or how many times primitive brains developed, brain-like structures were present in the ancient fish-like creatures that were ancestors of the vertebrates. The "arms race" that ensued as more and more sophisticated information-processing devices were developed led to many of the structures that are found in the human brain today, including the basal ganglia (which control patterns of movements) and the optic tectum (which is involved in the tracking of moving objects). Ultimately, as a direct result of evolutionary pressure, this "arms race" led to more and more intelligent brains. "Intelligence," Darwin wrote in *On the Origin of Species*, "is based on how efficient a species became at doing the things they need to survive."

About 200 million years ago, the descendants of the early vertebrates moved onto land and evolved further. One of the evolutionary branches eventually led to the mammals. Early mammals had a small neocortex, a part of the brain responsible for the complexity and flexibility of mammalian behavior. The brain size of mammals increased as they struggled to contend with the dinosaurs. Increases in brain size enabled mammals to improve their senses of smell, vision, and touch. Some of the mammals that survived the extinction of the dinosaurs took to the trees; they were the ancestors of the primates. Good eyesight, which helped them in that complex environment, led to an expansion of the visual part of the neocortex. Primates also developed bigger and more complex brains that enabled them to integrate and process the information reaching them and to plan and control their actions on the basis of that information.

The apes that lived some 14 million years ago in Africa were, therefore, smart, but probably not much smarter than their non-human descendants—orangutans, gorillas, and chimpanzees. Humans, however, evolved rapidly and in a different way. We don't know exactly what led to the fast development of the human brain, which began around 2.8 million years ago with the appearance of *Homo habilis* in Africa. It is likely that some mutation, or some small set of mutations, led to a growth of the brain that made the various species of the genus *Homo* smart enough to use technology and to develop language. The most salient development that occurred between earlier species and members of the genus *Homo* was the increase in cranial capacity from about 450 cubic centimeters in preceding species to 600 cubic centimeters in *Homo habilis* (Dawkins 2010).

Within the genus *Homo*, cranial capacity again doubled from *Homo habilis* to *Homo heidelbergensis* roughly 600,000 years ago, leading to a brain size comparable to that of *Homo sapiens*—roughly 1,200 cubic centimeters. Many different species of *Homo* have co-existed in the intervening millennia, including *Homo neanderthalensis*, which went extinct roughly 40,000 years ago (Higham et al. 2014), and *Homo floresiensis*, which lasted until about 12,000 years ago (Harari 2014).

By what may happen to be a historical coincidence, we humans have been, for twelve millennia, the only representatives of the genus *Homo* on Earth—a situation that may well be singular since the appearance of the genus nearly 3 million years ago. *Homo sapiens'* unique abilities to organize in large societies and to change its environment probably have been more responsible for this unique situation than any other factor (Harari 2014).

All the developments mentioned above led to the modern human brain, the most complex information-processing machine known to us. The modern human brain appeared in Africa about 200,000 years ago with the rise of *Homo sapiens*. Understanding how such a brain works is one of the most interesting challenges ever posed to science. But before we address this challenge, we must take a detour to survey the role computers have played in our understanding of complex biological systems and how computers can be used to model the human brain.

7 Biology Meets Computation

If the twentieth century was the century of physics, the twenty-first may well become the century of biology. One may argue that biology is a science with an already long and rich history and that the twenty-first century will be a century of many technologies. Still, the parallel has its merits. Biology will certainly be one the defining technologies of the twenty-first century, and, as happened with physics in the preceding century, exciting advances took place in biology in the first years of the new century. Much as physics changed in 1905 with the publication of Einstein's influential articles, biology changed in 2001 with the publication, by two competing teams, of two drafts of the whole sequence of the human genome.

Sequencing the Genome

Although DNA's structure and its role in genetics were known from the seminal work of Watson and Crick in the early 1950s, for about twenty years there was no technique that could be used to decode the sequence of bases in the genome of any living being. However, it was clear from the beginning that decoding the human genome would lead to new methods that could be used to diagnose and treat hereditary diseases, and to address many other conditions that depend on the genetic makeup of each individual.

The sequencing of the human genome resulted from the research effort developed in the decades that followed the discovery of DNA's structure. The interest in determining the sequence of the human genome ultimately led to the creation of the Human Genome Project, headed by James Watson, in 1990. That project's main goals were to determine the sequence of chemical base pairs that make up human DNA and to identify and map the tens of thousands of genes in the human genome. Originally expected to cost $3 billion and to take 15 years, the project was to be carried out by an international consortium that included geneticists in the United States, China, France, Germany, Japan, and the United Kingdom.

In 1998, a similar, but privately funded, project was launched by J. Craig Venter and his firm, Celera Genomics. The effort was intended to proceed at a much faster pace and at only a fraction of the cost of the Human Genome Project. The two projects used different approaches and reached the goal of obtaining a first draft of the human genome almost simultaneously in 2001.

The exponentially accelerating technological developments in sequencing technology that led to the sequencing of the human genome didn't stop in 2001, and aren't paralleled in any other field, even in the explosively growing field of digital circuit design. The rapidly advancing pace of sequencing technologies led to an explosion of genome-sequencing projects, and the early years of the twenty-first century saw a very rapid rise in the number of species with completely sequenced genomes.

The technological developments of sequencing technologies have outpaced even the very rapid advances of microelectronics. When the Human Genome Project started, the average cost for sequencing DNA was $10 per base pair. By 2000 the cost had fallen to one cent per base pair, and the improvements have continued at an exponential pace ever since. (See figure 7.1.) In 2015 the raw cost stood at roughly 1/100,000 of a cent per base pair, and it continues to fall, though more slowly than before (Wetterstrand 2015). The budget for the Human Genome Project was $3 billion (approximately $1 per base pair). In 2007, Watson's own genome was sequenced using a new technology, at a cost of $2 million. This may seem expensive, but it was less than a thousandth as expensive as the original Human Genome Project. As of this writing, the cost of sequencing a human genome is approaching the long-sought target of $1,000. The cost per DNA base pair has fallen by a factor of 200,000 in 15 years, which is a reduction by a factor of more than 2 every year, vastly outpacing the reduction in costs that results from Moore's Law, which has reduced costs by a factor of 2 every two years.

Sequencing technology had a slow start in the years that followed the discovery of DNA, despite the growing interest in methods that could be used to determine the specific sequence of bases in a specific strand of DNA. About twenty years passed between the discovery of the DNA double

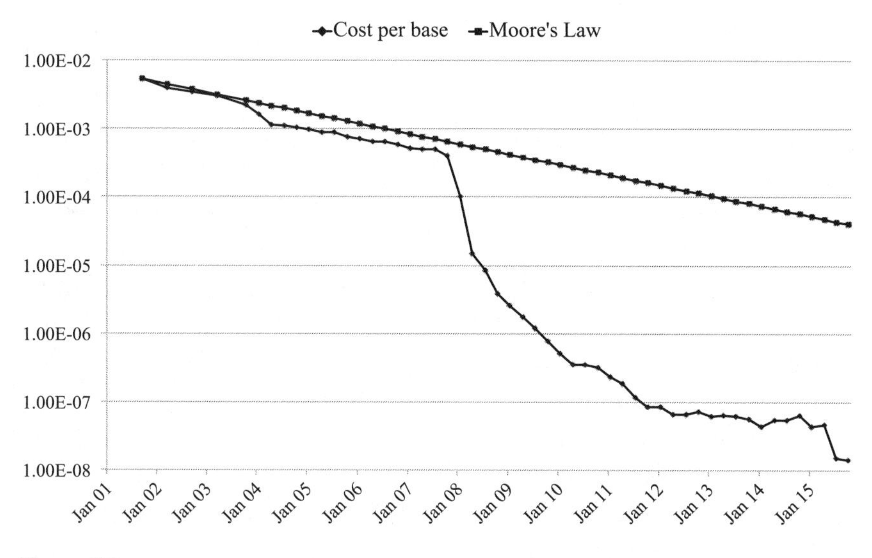

Figure 7.1
A graph comparing sequencing costs with Moore's Law, based on data compiled by the National Human Genome Research Institute. Values are in US dollars per DNA base.

helix and the invention of the first practical methods that could be used to determine the actual bases in a DNA sequence. Since it is not possible to directly observe the sequence of bases in a DNA chain, it is necessary to resort to indirect methods. The most practical of these first methods was the chain-terminator method (Sanger, Nicklen, and Coulson 1977), still known today as Sanger sequencing. The principle behind that method is the use of specific molecules that act to terminate the growth of a chain of DNA in a bath of deoxynucleotides, the essential constituents of DNA. DNA sequences are then separated by size, which makes it possible to read the sequence of base pairs. The method starts by obtaining single-stranded DNA by a process called *denaturation*. Denaturation is achieved by heating DNA until the double-stranded helix separates into two strands. The single-stranded DNA is then used as a template to create new double-stranded DNA, which is obtained by adding, in four physically separate reactions, the four standard DNA deoxynucleotides, A, C, T, and G. Together with the standard deoxynucleotides, each reaction also contains a solution of one modified deoxynucleotide, a dideoxynucleotide that has the property of stopping the growth of the double-stranded DNA when it is incorporated. Each of these chain terminators attaches to one specific base (A, C, T, or G) on the single-stranded DNA and stops the DNA double chain from growing further. DNA fragments of different sizes can then be separated by making them flow through a gel through which fragments of different sizes flow at different speeds. From the positions of the fragments in the gel, the specific sequence of bases in the original DNA fragment can be read.

The Sanger sequencing technique enabled researchers to deal with small DNA sequences and was first applied to the study of small biological subsystems. In-depth study of specific genes or other small DNA sequences led to an increasing understanding of cellular phenomena.

For a number of decades, the ability to sequence the complete genome of any organism was beyond the reach of existing technologies. As the technology evolved, sequencing of longer and longer sequences led to the first sequencing of a complete gene in 1972 and to the first complete sequencing of an organism (the 3,569-base DNA sequence of the bacteriophage MS2) in 1976 (Fiers et al. 1976). This was, however, still very far from sequencing any complex organism. (Complex organisms have billions of bases in their genomes.)

The first realistic proposal for sequencing the human genome appeared in 1985. However, not until 1988 did the Human Genome Project make it clear that such an ambitious objective was within reach. The selection of model organisms with increasing degrees of complexity enabled not only the progressive development of the technologies necessary for the sequencing of the human genome, but also the progressive development of research in genomics. The model organisms selected to be sequenced included *Hemophilus influenza* (with 1.8 mega bases, or Mb), in 1995, *Escherichia coli* (with 5 Mb), in 1996, *Saccharomyces cerevisiae* (baker's yeast, with 12 Mb), in 1996, *Caenorhabditis elegans* (a roundworm, with 100 Mb), in 1998, *Arabidopsis thaliana* (a small flowering plant, with 125 Mb), in 2000, and *Drosophila melanogaster* (the fruit fly, with 200 Mb), in 2000. The final steps of the process resulted in the simultaneous publication, in 2001, of two papers on the human genome in the journals *Science* and *Nature* (Lander et al. 2001; Venter et al. 2001). The methods that had been used in the two studies were different but not entirely independent.

The approach used by the Human Genome Project to sequence the human genome was originally the same as the one used for the *S. cerevisiae* and *C. elegans* genomes. The technology available at the time the project started was the BAC-to-BAC method. A bacterial artificial chromosome (BAC) is a human-made DNA sequence used for transforming and cloning DNA sequences in bacteria, usually *E. coli*. The size of a BAC is usually between 150 and 350 kilobases (kb).

The BAC-to-BAC method starts by creating a rough physical map of the genome. Constructing this map requires segmenting the chromosomes into large pieces and figuring out the order of these big segments of DNA before sequencing the individual fragments. To achieve this, several copies of the genome are cut, in random places, into pieces about 150 kb long. Each of these fragments is inserted into a BAC. The collection of BACs containing the pieces of the human genome is called a *BAC library*. Segments of these pieces are "fingerprinted" to give each of them a unique identification tag that can be used to determine the order of the fragments. This is done by cutting each BAC fragment with a single enzyme and finding common sequences in overlapping fragments that determine the location of each BAC along the chromosome. The BACs overlap and have these markers every 100,000 bases or so, and they can therefore be used to create a physical map of each chromosome. Each BAC is then sliced randomly into smaller pieces (between 1 and 2 kb) and sequenced, typically about 500 base pairs from each end of the fragment. The millions of sequences which are the result of this process are then assembled (by using an algorithm that looks for overlapping sequences) and placed in the right place in the genome (by using the physical map generated).

The "shotgun" sequencing method, proposed and used by Celera, avoids the need for a physical map. Copies of the genome are randomly sliced into pieces between 2 and 10 kb long. Each of these fragments is inserted into a plasmid (a piece of DNA that can replicate in bacteria). The so-called plasmid libraries are then sequenced, roughly 500 base pairs from each end, to obtain the sequences of millions of fragments. Putting all these sequences together required the use of sophisticated computer algorithms.

Further developments in sequencing led to the many technologies that exist today, most of them based on different biochemical principles. Present-day sequencers generate millions of DNA reads in each run. A run takes two or three days and generates a volume of data that now approaches a terabyte (10^{12} bytes)—more than the total amount of data generated in the course of the entire Human Genome Project.

Each so-called read contains the sequence of a small section of the DNA being analyzed. These sections range in size from a few dozen base pairs to several hundred, depending on the technology. Competing technologies are now on the market (Schuster 2007; Metzker 2010), the most common being pyrosequencing (Roche 454), sequencing by ligation (SOLiD), sequencing by synthesis (Illumina), and Ion Torrent. Sanger sequencing still remains the technology that provides longer reads and less errors per position, but its throughput is much lower than the throughput of next-generation sequencers, and thus it is much more expensive to use.

Computer scientists became interested in algorithms for the manipulation of biological sequences when it was recognized that, in view of the number and the lengths of DNA sequences, it is far from easy to mount genomes from reads or even to find similar sequences in the middle of large volumes of data.

Because all living beings share common ancestors, there are many similarities in the DNA of different organisms. Being able to look for these similarities is very useful because the role of a particular gene in one organism is likely to be similar to the role of a similar gene in another organism. The development of algorithms with which biological sequences (particularly DNA sequences) could be processed efficiently led to the appearance of a new scientific field called *bioinformatics*. Especially useful for sequencing the human genome were algorithms that could be used to manipulate very long sequences of symbols, also known as *strings*. Although manipulating small strings is relatively easy for computers, the task becomes harder when the strings have millions of symbols, as DNA sequences do. Efficient manipulation of strings became one of the most important fields in computer science. Efficient algorithms for manipulating strings are important both in bioinformatics and in Web technology (Baeza-Yates and Ribeiro-Neto 1999; Gusfield 1997).

A particularly successful tool for bioinformatic analysis was BLAST, a program that looks for exact or approximate matches between biological sequences (Altschul et al. 1990). BLAST is still widely used to search databases for sequences that closely match DNA sequences or protein sequences of interest. Researchers in bioinformatics developed many other algorithms for the manipulation of DNA strings, some of them designed specially to perform the assembly of the millions of fragments generated by the sequencers into the sequence of each chromosome (Myers et al. 2000).

Even though BAC-to-BAC sequencing and shotgun sequencing look similar, both involving assembly of small sequence fragments into the complete sequences of the chromosomes, there is a profound difference between the two approaches. In BAC-to-BAC sequencing, the physical map is known, and finding a sequence that corresponds to the fragments is relatively easy because we know approximately where the segment belongs. It's a bit like solving a puzzle for which the approximate location of each piece is roughly known, perhaps because of its color. However, in the case of the human genome creating the BAC library and deriving the physical map was a long and complex process that took many years to perfect and to carry out. In shotgun sequencing this process is much faster and less demanding, since the long DNA sequences are immediately broken into smaller pieces and no physical map is constructed. On the other hand, assembling millions of fragments without knowing even approximately where they are in the genome is a much harder computational problem, much like solving a puzzle all of whose pieces are of the same color. Significant advances in algorithms and in computer technology were necessary before researchers were able to develop programs that could perform this job efficiently enough.

One algorithm that has been used to construct DNA sequences from DNA reads is based on the construction of a graph (called a *de Bruijn graph*) from the reads. In such a graph, each node corresponds to one specific fixed-length segment of k bases, and two nodes in the graph are connected if there is a read with $k + 1$ bases composed of the overlapping of two different k-base segments.

For example, suppose you are given the following set of four base reads: TAGG, AGGC, AGGA, GGAA, GAAT, GCGA, GGCG, CGAG, AATA, GAGG, ATAG, TAGT. What is the shortest DNA sequence that could have generated all these reads by being cut in many different places? The problem can be solved by drawing the graph shown here in figure 7.2, in which each node corresponds to a triplet of bases and each edge corresponds to one read. Finding the smallest DNA sequence that contains all these reads is then equivalent to finding a Eulerian path in this directed graph—that is, a path in the graph starting in one node and going through each edge exactly once (Pevzner, Tang, and Waterman 2001). As we saw in chapter 4, for undirected graphs there exists such a path if, and only if, exactly two vertices have an odd degree. For directed graphs, Euler's theorem must be

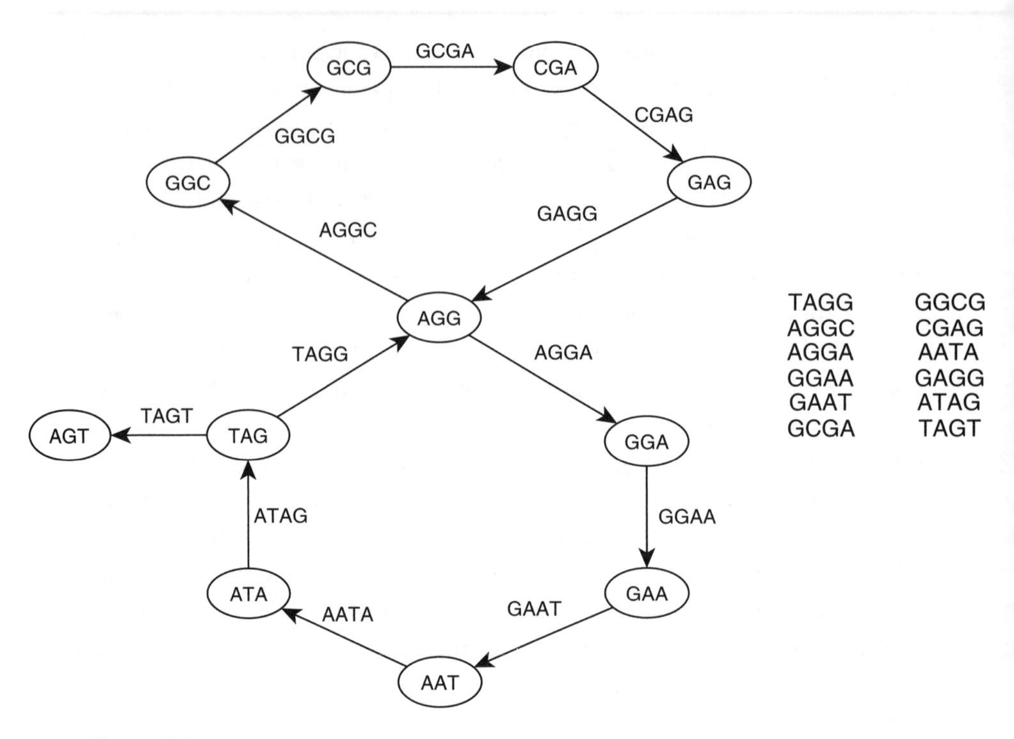

Figure 7.2
DNA sequencing from reads, using Eulerian paths. The reads on the right are used to label the edges on the graph. Two nodes are connected, if they have labels contained in the read that is used to label the edge that connects them.

changed slightly, to take into account the fact that edges can be traversed in only one direction. A directed graph has a Eulerian path if every vertex has the same in degree and out degree, except for the starting and ending vertices (which must have, respectively, an out degree equal to the in degree plus 1 and vice versa). In this graph, all vertices have the in degree equal to the out degree except for vertices TAG and AGT, which satisfy the condition for the starting and ending vertices respectively. The Eulerian path, therefore, starts in node TAG, goes through node AGG and around the nodes in the topmost loop, returns to node AGG, and finishes in node AGT after going through the nodes in the bottom loop. This path gives TAGGCGAGGAATAGT, the shortest DNA sequence containing all the given reads.

It is also possible to formulate this problem by finding a Hamiltonian path in a graph in which each read is represented by a node and the overlap between reads is represented by an edge (Compeau, Pevzner, and Tesler 2011). That approach, which was indeed used in the Human Genome Project (Lander et al. 2001; Venter et al. 2001), is computationally much more intensive and less efficient, because the Hamiltonian path problem is an NP-complete problem. In practice, mounting a genome using de Bruijn graphs is more complex than it appears to be in figure 7.2 because of errors in the sequences, missing sequences, highly repetitive DNA regions, and several other factors.

The race to sequence the human genome, which ended in a near tie, made it clear that computers and algorithms were highly useful in the study of biology. In fact, assembling the human genome wouldn't have been possible without the use of these algorithms, and they became the most important component of the technologies that were developed for the purpose. The field of bioinformatics was born with these projects. Since then it has became a large enterprise involving many research groups around the world.

Today, many thousands of researchers and professionals work in bioinformatics, applying computers to problems in biology and medicine. Bioinformatics, (also called *computational biology*) is now a vast field with many sub-fields. In fact, manipulating DNA sequences is only one of the many tasks routinely performed by computers, helping us along the long path that will lead to a more complete understanding of biological systems. Researchers and professionals in bioinformatics are a diverse lot and apply their skills to many different challenges, including the modeling and simulation of many types of biological networks. These networks—abstract models that are used to study the behaviors of biological systems—are essential tools in our quest to understand nature.

Biological Networks

A multi-cellular organism consists of many millions of millions of cells. The human body contains more than 30 trillion cells (Bianconi et al. 2013), a number so large as to defy the imagination.

Although DNA contains the genetic information of an organism, there are many more components inside a living cell. Recently developed technologies enable researchers to obtain vast amounts of data related to the concentrations of various chemicals inside cells. These measurements are critically important for researchers who want to understand cellular behavior, in particular, and biological systems, in general. The advances in genomics have been paralleled by advances in other areas of biotechnology, and the amounts of data now being collected by sequencing centers and biological research institutes are growing at an ever-increasing rate.

To make sense of these tremendous amounts of data, computer models of biological processes have been used to study the biological systems that provided the data. Although there are many types of biological models, the most widespread and useful models are biological networks. In general, it is useful to view complex biological systems as a set of networks in which the various components of a system interact. Networks are usually represented as graphs in which each node is a component in the network and the edges between the nodes represent interactions between components.

Many types of biological networks have been used to model biological systems, and each type of network has many different sub-types. Here I will use a small number of types of biological networks that are representative to illustrate how models of biological networks are used to study, simulate, and understand biological systems.

In general, nodes of biological networks need not correspond to actual physically separate locations. The network represents, in this case, a useful abstraction, but there is no one-to-one correspondence between a node in a network and a physical location, as there is in a road network or in an electrical network. Of the four types of networks I will discuss, only in neuronal networks is there direct correspondence between a node in the network and a specific location in space, and even in that case the correspondence isn't perfect.

The four types of biological networks I will describe here are metabolic networks, protein-protein interaction networks, gene regulatory networks, and neuronal networks. I will use the term *neuronal networks* to refer to networks that model biological neurons, in order to distinguish them from neural networks (which were discussed in chapter 5 as a machine learning technique). Even though other types of biological networks are used to model biological systems, these four types constitute a large fraction of the models used to study the behavior of cells, systems, and even whole organisms.

Metabolic networks are used to model the interactions among chemical compounds of a living cell, which are connected by biochemical reactions that convert one set of compounds into another. Metabolic networks are commonly used to model the dynamics of the concentrations of various chemical compounds inside a cell. These networks describe the structure and the dynamics of the chemical reactions of metabolism (also known as the metabolic pathways), as well as the mechanisms used to regulate the interactions between compounds. The reactions are catalyzed by enzymes, which are selective catalysts, accelerating the rates of the metabolic reactions that are involved in most biological processes, from the digestion of nutrients to the synthesis of DNA. Although there are other types of enzymes, most enzymes are proteins and are therefore encoded in the DNA. When the complete genome of an organism is known, it is possible to reconstruct partially or entirely the network of biochemical reactions in a cell, and to derive mathematical expressions that describe the dynamics of the concentration of each compound.

Protein-protein interaction networks are used very extensively to model the dynamics of a cell or a system of cells. Proteins, encoded by the genes, interact to perform a very large fraction of the functions in a cell. Many essential processes in a cell are carried out by molecular machines built from a number of protein components. The proteins interact as a result of biochemical events, but mainly because of electrostatic forces. In fact, the quaternary structure of proteins, to which I alluded in chapter 6, is the arrangement of several folded proteins into a multi-protein complex. The study of protein-protein interaction networks is fundamental to an understanding of many processes inside a living cell and of the interaction between different cells in an organism.

Gene regulatory networks are used to model the processes that lead to differentiated activity of the genes in different environments. Even though all cells in a multi-cellular organism (except gametes, the cells directly involved in sexual reproduction) have the same DNA, there are many different kinds of cells in a complex organism and many different states for a given cell type. The process of cellular differentiation creates a more specialized cell from a less specialized cell type, usually starting from stem cells, in adult organisms. Cell differentiation leads to cells with different sizes, shapes, and functions and is controlled mainly by changes in gene expression. In different cell types (and even in different cells of the same type in different states) different genes are expressed and thus, different cells can have very different characteristics, despite sharing the same genome. Even a specific cell can be in many different states, depending on the genes expressed at a given time. The activity of a gene is controlled by a number of pre-transcription and post-transcription regulatory mechanisms. The simplest mechanism is related to the start of the transcription. A gene can be transcribed (generating mRNA that will then be translated into a protein) only if a number of proteins (called *transcription factors*) are attached to a region near the start of the coding region of the gene: the promoter region. If for some reason the transcription factors don't attach to the promoter region, the gene is not transcribed and the protein that it generates is not created. Gene regulatory networks, which model this mechanism, are, in fact, DNA-protein interaction networks. Each node in this network is either a transcription factor (a protein) or a gene (a section of DNA). Gene regulatory networks can be used to study cell behavior, in general, and cell differentiation, in particular. Understanding the behavior and the dynamics of gene regulatory networks is an essential step toward understanding the behavior of complex biological systems.

Neuronal networks are one type of cell interaction networks extensively used to study brain processes. Although different types of cell interaction networks are commonly used, the most interesting type is neuron-to-neuron interaction networks, the so-called neuronal networks. The nervous system of a complex organism consists of interconnected neural cells, supported by other types of cells. Modeling this network of neurons is a fundamental step toward a more profound understanding of the behavior of nervous systems. Each node in a neuronal network corresponds to one nerve cell, and an edge corresponds to a connection between neurons. The activity level of each neuron and the connections (excitatory or inhibitory) between neurons lead to patterns of neuron excitation in the nervous system that correspond to specific brain behaviors. In the central nervous systems of humans and in those of our close evolutionary parents, the primates, these patterns of behavior lead to intelligent and presumably conscious behavior. The study of neuronal networks is, therefore, of fundamental importance to the understanding of brain behavior.

As we saw in chapter 5, simplified mathematical models of neurons have been used to create artificial neural networks. As I noted, artificial neural networks capture some characteristics of networks of neurons, even though they are not faithful models of biological systems. Networks of these artificial neurons—artificial neural networks— have been used extensively both as models of parts of the brain and as mechanisms to perform complex computational tasks requiring adaptation and learning. However, in general, artificial neural networks are not used to model actual cell behavior in the brain. Modeling actual cell behaviors in living brains is performed using more sophisticated models of neurons, which will be described in chapter 8.

Emulating Life

The development of accurate models for biological systems will enable us to simulate *in silico* (that is, by means of a computer) the behavior of actual biological systems. Simulation of the dynamic behavior of systems is common in many other fields of science. Physicists simulate the behavior of atoms, molecules, mechanical bodies, planets, stars, star systems, and even whole galaxies. Engineers routinely use simulation to predict the behaviors of electronic circuits, engines, chemical reactors, and many others kinds of systems. Weather forecasters simulate the dynamics of the atmosphere, the oceans, and the land to predict, with ever-increasing accuracy, tomorrow's weather. Economists simulate the behavior of economic systems and, with some limited success, of whole economies.

In computer science the word *simulation* has a very special meaning—it usually refers to the simulation of one computer by another. In theoretical computer science, if one computing engine can simulate another then they are, in some sense, equivalent. As I noted in chapter 4, in this situation I will use the word *emulate* instead of *simulate*. We have already encountered universal Turing machines, which can emulate any other Turing machine, and the idea that a universal Turing machine can emulate any existing computer, if only slowly. In more practical terms, it has become common to use the concept of *virtualization*, in which one computer, complete with its operating system and its programs, is emulated by a program, running in some other computer. This other computer may, itself, be also just an emulation running in another computer. At some point this regression must end, of course, and some actual physical computer must exist, but there are no conceptual limits to this virtualization of computation.

The ability to simulate the behavior of a system requires the existence of an abstract computation mechanism, either analog or digital. Although analog simulators have been used in the past (as we saw in chapter 2), and although they continue to be used to some extent, the development of digital computers made it possible to simulate very complex systems flexibly and efficiently. Being able to simulate a complex system is important for a number of reasons, the most important of which are the ability to adjust the time scale, the ability to observe variables of interest, the ability to understand the behavior of the system, and the ability to model systems that don't exist physically or that are not accessible.

The time scales involved in the behavior of actual physical systems may make it impossible to observe events in those systems, either because the events happen too slowly (as in a collision of two galaxies) or because the events happen too fast (as in molecular dynamics). In a simulation, the time scales can be compressed or expanded in order to make observation of the dynamics of interest possible. A movie of two galaxies colliding lasting only a minute can compress an event that, in reality, would last many hundreds of millions of years. Similarly, the time scale of molecular dynamics can be slowed down by a factor of many millions to enable researchers to understand the dynamic behavior of atoms and molecules in events that happen in nanoseconds.

Simulation also enables scientists and engineers to observe any quantity of interest (e.g., the behavior of a part of a transistor in an integrated circuit)—something that, in many cases, is not possible in the actual physical system. Obviously one can't measure the temperature inside a star, or even the temperature at the center of the Earth, but simulations of these systems can tell us, with great certainty, that the temperature at the center of the Sun is roughly 16 million kelvin and that the temperature at the center of the Earth is approximately 6,000 kelvin. In a simulator that is detailed enough, any variable of interest can be modeled and its value monitored.

The ability to adjust the time scales and to monitor any variables internal to the system gives the modeler a unique insight into the behavior of the system, ultimately leading to the possibility of a deep understanding that cannot otherwise be obtained. By simulating the behavior of groups of atoms that form a protein, researchers can understand why those proteins perform the functions they were evolved to perform. This deep understanding is relatively simple in some cases and more complex (or even impossible) in others, but simulations always provide important insights into the behavior of the systems—insights that otherwise would be impossible to obtain.

The fourth and final reason why the ability to simulate a system is important is that it makes it possible to predict the behavior of systems that have not yet come into existence. When a mechanical engineer simulates the aerodynamic behavior of a new airplane, the simulation is used to inform design decisions that influence the design of the plane before it is built. Similarly, designers of computer chips simulate the chips before they are fabricated. In both cases, simulation saves time and money.

Simulation of biological systems is a comparatively new entrant in the realm of simulation technology. Until recently, our understanding of biological systems in general, and of cells in particular, was so limited that we weren't able to build useful simulators. Before the last two decades of the twentieth century, the dynamics of cell behavior were so poorly understood that the knowledge needed to construct a working model was simply not available. The understanding of the mechanisms that control the behavior of cells, from the processes that create proteins to the many mechanisms of cell regulation, opened the door to the possibility of detailed simulation of cells and of cell systems.

The dynamics of the behavior of a single cell are very complex, and the detailed simulation of a cell at the atomic level is not yet possible with existing computers. It may, indeed, never be possible or economically feasible. However, this doesn't mean that very accurate simulation of cells and of cell systems cannot be performed. The keys to the solution are *abstraction* and *hierarchical multi-level simulation,* concepts extensively used in the simulation of many complex systems, including, in particular, very-large-scale integrated (VLSI) circuits.

The idea behind abstraction is that, once the behavior of a part of a system is well understood, its behavior can be modeled at a higher level. An abstract model of that part of the system is then obtained, and that model can be used in a simulation, performed at a higher level, leading to hierarchical multi-level simulation, which is very precise and yet very efficient. For instance, it is common to derive an electrical-level model for a single transistor. The transistor itself is composed of many millions or billions of atoms, and detailed simulation, at the atom and electron level, of a small section of the transistor may be needed to characterize the electrical properties of the device. This small section of the transistor is usually simulated at a very detailed physical level. The atoms in the crystalline structure of the semiconductor material are taken into consideration in this simulation, as are the atoms of impurities present in the crystal. The behavior of these structures, and that of the electrons present in the electron clouds of these atoms, which are subject to electric fields, is simulated in great detail, making it possible to characterize the behavior of this small section of the transistor. Once the behavior of a small section of the transistor is well understood and well characterized, an electrical-level model for the complete transistor can be derived and used to model very accurately the behavior of the device, even though it doesn't model each individual atom or electron. This electrical-level model describes how much current flows through a transistor when a voltage difference is applied at its terminals; it also describes other characteristics of the transistor. It is precise enough to model the behavior of a transistor in a circuit, yet it is many millions times faster than the simulation at the atomic level that was performed to obtain the model.

Thousands or even millions of transistors can then be effectively simulated using these models for the transistors, even if the full simulation at the electron level and at the atom level remains impossible. In practice, complete and detailed simulation of millions (or even billions) of transistors in an integrated circuit is usually not carried out. Systems containing hundreds or thousands of transistors are characterized, using these electrical-level models and more abstract models of these systems are developed and used, in simulations performed at a higher level of abstraction. For instance, the arithmetic and logic unit (ALU) that is at the core of a computer consists of a few thousand transistors, but they can be characterized solely in terms of how much time and power the ALU requires to compute a result, depending on the type of operation and the types of operands. Such a model is then used to simulate the behavior of the integrated circuit when the unit is connected to memory and to other units. High-level models for the memory and for the other units are also used, making it practicable to perform the simulation of the behavior of billions of transistors, each consisting of billions of atoms. If one considers other parts in the circuit that are also being implicitly simulated, such as the wires and the insulators, this amounts to complete simulation of a system with many billions of trillions of atoms.

Techniques for the efficient simulation of very complex biological systems also use this approach, even though they are, as of now, much less well developed than the techniques used in the simulation of integrated circuits. Researchers are working intensively on the development of models for the interaction between molecules in biological systems—especially the interactions among DNA, proteins, and metabolites. Models for the interaction between many different types of molecules can then be used in simulations at higher levels of abstraction. These higher levels of abstraction include the various types of biological networks that were discussed in the preceding section. For instance, simulation of a gene regulatory network can be used to derive the dynamics of the concentration of a particular protein when a cell responds to specific aggression by a chemical compound. The gene regulatory network represents a model at a high level of abstraction that looks only at the concentrations of a set of proteins and other molecules inside a cell. However, that model can be used to simulate the phenomena that occur when a specific transcription factor attaches to the DNA near the region where the gene is encoded. Modeling the attachment between a transcription factor (or a set of transcription factors) and a section of the DNA is a very complex task that may require the extensive simulation of the molecular dynamics involved in the attachment. In fact, the mechanism is so complex that it is not yet well understood, and only incomplete models exist. However, extensive research continues, and good detailed models for this process will eventually emerge. It will then be practicable to use those detailed models to obtain a high-level model that will relate the concentration of the transcription factors with the level of transcription of the gene, and to use the higher-level models in the gene regulatory network model to simulate the dynamics of the concentrations of all the molecules involved in the process, which will lead to a very efficient and detailed simulation of a system that involves many trillions of atoms.

By interconnecting the various types of networks used to model different aspects of biological systems and using more and more abstract levels of modeling, it may eventually become possible to model large systems of cells, complete organs, and even whole organisms. Modeling and simulating all the processes in an organism will not, in general, be of interest. In most cases, researchers will be interested in modeling and understanding specific processes, or pathways, in order to understand how a given dysfunction or disease can be tackled. Many of the problems of interest that require the simulation of complex biological systems are related to illnesses and other debilitating conditions. In fact, every existing complex disease requires the work of many thousands of researchers in order to put together models that can be used to understand the mechanisms of disease and to investigate potential treatments. Many of these treatments involve designing specific molecules that will interfere with the behavior of the biological networks that lead to the disease. In many types of cancer, the protein and gene regulatory networks controlling cell division and cell multiplication are usually the targets of study, with the aim of understanding how uncontrolled cell division can be stopped. Understanding neurodegenerative diseases that progressively destroy or damage neuron cells requires accurate models of neuronal networks and also of the molecular mechanisms that control the behavior of each neuron. Aging isn't a disease, as such, but its effects can be delayed (or perhaps averted) if better models for the mechanisms of cellular division and for the behavior of networks of neurons become available.

Partial models of biological organisms will therefore be used to understand and cure many illnesses and conditions that affect humanity. The use of these models will not, in general, raise complex ethical problems, since they will be *in silico* models of parts of organisms and will be used to create therapies. In most cases, they will also reduce the need for animal experiments to study the effects of new molecules, at least in some phases of the research. However, the ability to perform a simulation of a complete organism raises interesting ethical questions because, if accurate and detailed enough, it will reproduce the behavior of the actual organism, and it would become, effectively, an emulation. Perhaps the most advanced effort to date to obtain the simulation of a complex organism is the OpenWorm project, which aims at simulating the entire roundworm *Caenorhabditis elegans*.

Digital Animals

The one-millimeter-long worm *Caenorhabditis elegans* has a long history in science as a result of its extensive used as a model for the study of simple multi-cellular organisms. It was, as was noted above, the first animal to have its genome sequenced, in 1998. But well before that, in 1963, Sydney Brenner proposed it as a model organism for the investigation of neural development in animals, an idea that would lead to Brenner's research at the MRC Laboratory for Molecular Biology in Cambridge, England, in the

1970s and the 1980s. In an effort that lasted more than twelve years, the complete structure of the brain of *C. elegans* was reverse engineered, leading to a diagram of the wiring of each neuron in this simple brain. The work was made somewhat simpler by the fact that the structure of the brain of this worm is completely encoded by the genome, leading to virtually identical brains in all the individuals that share a particular genome.

The painstaking effort of reverse engineering a worm brain included slicing several worm brains very thinly, taking about 8,000 photos of the slices with an electron microscope, and connecting each neuron section of each slice to the corresponding neuron section in the neighbor slices (mostly by hand). The complete wiring diagram of the 302 neurons and the roughly 7,000 synapses that constitute the brain of *C. elegans* was described in minute detail in a 340-page article titled "The Structure of the Nervous System of the Nematode *Caenorhabditis elegans*" (White et al. 1986). The running head of the article was shorter and more expressive: "The Mind of a Worm." This detailed wiring diagram could, in principle, be used to create a very detailed simulation model with a behavior that would mimic the behavior of the actual worm. In fact, the OpenWorm project aims at constructing a complete model, not only of the 302 neurons and the 95 muscle cells, but also of the remaining 1,000 cells in each worm (more exactly, 959 somatic cell plus about 2,000 germ cells in the hermaphrodites and 1,031 cells in the males).

The OpenWorm project, begun in 2011, has to overcome significant challenges to obtain a working model of the worm. The wiring structure of the brain, as obtained by Brenner's team, isn't entirely sufficient to define its actual behavior. Much more information is required, such as detailed models for each connection between neurons (the synapses), dynamical models for the neurons, and additional information about other variables that influence the brain's behavior. However, none of this information is, in principle, impossible to obtain through a combination of further research in cell behavior, advanced instrumentation techniques (which I will address in chapter 9), and fine tuning of the models' parameters.

The OpenWorm project brings together coders (who will write the complete simulator using a set of programming languages) and scientists working in many fields, including genetics, neuron modeling, fluid dynamics, and chemical diffusion. The objective is to have a complete simulator of all the electrical activity in all the muscles and neurons, and an integrated simulation environment that will model body movement and physical forces within the worm and the worm's interaction with its environment.

The interesting point for the purpose of the present discussion is that, in principle, the OpenWorm project will result in the ability to simulate the complete behavior of a fairly complex animal by emulating the cellular processes in a computer. Let us accept, for the moment, the reasonable hypothesis that this project will succeed in creating a detailed simulation of *C. elegans*. If its behavior, in the presence of a variety of stimuli, is indistinguishable from a real-world version of the worm, one has to accept that the program which simulates the worm has passed, in a very restricted and specific way, a sort of Turing Test: A program, running in a computer, simulates a worm well enough to fool an observer. One may argue that a simulation of a worm will never fool an observer as passing for the real thing. This is, however, a simplistic view. If one has a good simulation of the worm, it is relatively trivial to use standard computer graphics technologies to render its behavior on a screen, in real time, in such a way that it will not be distinguishable from the behavior of a real worm.

In reality, no one will care much if a very detailed simulation of a worm is or is not indistinguishable from the real worm. This possibility, however, opens the door to a much more challenging possibility: What if, with the evolution of science and technology, we were to become able to simulate other, more complex animals in equivalent detail? What if we could simulate, in this way, a mouse, a cat, or even a monkey, or at least significant parts of those animals? And if we can simulate a monkey, what stops us from eventually being able to simulate a human? After all, primates and humans are fairly close in biological complexity. The most challenging differences are in the brain, which is also the most interesting organ to simulate. Even though the human brain has on the order of 100 billion neurons and the brain of a chimpanzee only 7 billion, this difference of one order of magnitude will probably not be a roadblock if we are ever able to reverse engineer and simulate a chimp brain. A skeptical reader probably will argue, at this point, that this is a far-fetched possibility, not very likely to happen any time soon. It is one thing to simulate a worm with 302 neurons and no more than 1,000 cells; it is a completely different thing to simulate a more complex animal, not to speak of a human being with a hundred billion neurons and many trillions of cells.

Nonetheless, we must keep in mind that such a possibility exists, and that simulating a complex animal is, in itself, a worthwhile goal that, if achieved, will change our ability to understand, manipulate, and design biological systems. At present this ability is still limited, but advances in synthetic biology may change that state of affairs.

Synthetic Biology

Advances in our understanding of biological systems have created the exciting possibility that we will one day be able to design new life forms from scratch. To do that, we will have to explore the vastness of the Library of Mendel in order to find the specific sequences of DNA that, when interpreted by the cellular machinery, will lead to the creation of viable new organisms. Synthetic biology is usually defined as the design and construction of biological devices and systems for useful purposes. The geneticist Wacław Szybalski may have been the first to use the term. In 1973, when asked during a panel discussion what he would like to be doing in the "somewhat non-foreseeable future," he replied as follows:

> Let me now comment on the question "what next." Up to now we are working on the descriptive phase of molecular biology. ... But the real challenge will start when we enter the synthetic phase of research in our field. We will then devise new control elements and add these new modules to the existing genomes or build up wholly new genomes. This would be a field with an unlimited expansion potential and hardly any limitations to building "new better control circuits" or ... finally other "synthetic" organisms, like a "new better mouse." ... I am not concerned that we will run out of exciting and novel ideas ... in the synthetic biology, in general.

Designing a new biological system is a complex enterprise. Unlike human-made electronic and mechanical systems, which are composed of modular components interconnected in a clear way, each with a different function, biological systems are complex contraptions in which each part interacts with all the other parts in complicated and unpredictable ways. The same protein, encoded in a gene in DNA, may have a function in the brain and another function, entirely different, in the skin. The same protein may fulfill many different functions in different cells, and at present we don't have enough knowledge to build, from scratch, an entirely new synthetic organism.

However, we have the technology to create specific sequences of DNA and to insert them into cells emptied of their own DNA. The machinery of those cells will interpret the inserted synthetic DNA to create new copies of whatever cell is encoded in the DNA. It is now possible to synthetize, from a given specification, DNA sequences with thousands of base pairs. Technological advances will make it possible to synthesize much longer DNA sequences at very reasonable costs.

The search for new organisms in the Library of Mendel is made easier by the fact that some biological parts have well-understood functions, which can be replicated by inserting the code for these parts in the designed DNA. The DNA parts used most often are the BioBrick plasmids invented by Tom Knight (2003). BioBricks are DNA sequences that encode for specific proteins, each of which has a well-defined function in a cell. BioBricks are stored and made available in a registry of standard biological parts and can be used by anyone interested in designing new biological systems. In fact, this "standard" for the design of biological systems has been extensively used by thousands of students around the world in biological system design competitions such as the iGEM competition. Other parts and methodologies, similar in spirit to the BioBricks, also exist and are widely available.

Existing organisms have already been "re-created" by synthesizing their DNA and introducing it in the machinery of a working cell. In 2003, a team from the J. Craig Venter Institute synthetized the genome of the bacteriophage Phi X 174, the first organism to have had its 5386 base genome sequenced. The team used the synthesized genome to assemble a living bacteriophage. Most significantly, in 2006, the same team constructed and patented a synthetic genome of *Mycoplasma laboratorium*, a novel minimal bacterium derived from the genome of *Mycoplasma genitalium*, with the aim of creating a viable synthetic organism. *Mycoplasma genitalium* was chosen because it was the known organism that exhibits the smallest number of genes. Later the research group decided to use another bacterium, *Mycoplasma mycoides*, and managed to transplant the synthesized genome of this bacterium into an existing cell of a different bacterium that had had its DNA removed. The new bacterium was reported to have been viable and to have replicated successfully.

These and other efforts haven't yet led to radically new species with characteristics and behaviors completely different from those of existing species. However, there is little doubt that such efforts will continue and will eventually lead to new technologies that will enable researchers to alter existing life forms. Altered life forms have the potential to solve many problems, in many different areas. In theory, newly designed bacteria could be used to produce hydrocarbons from carbon dioxide and sunlight, to clean up oil spills, to fight global warming, and to perform many, many other tasks.

If designing viable new unicellular organisms from scratch is still many years in the future, designing complex multi-cellular eukaryotes is even farther off. The developmental processes that lead to the creation of different cells types and body designs are very complex and depend in very complicated ways on the DNA sequence. Until we have a much more complete understanding of the way biological networks work in living organisms, we may be able to make small "tweaks" and adjustments to existing life forms, but we will not be able to design entirely new species from scratch. Unicorns and centaurs will not appear in our zoos any time soon. In a way, synthetic biology will be, for many years to come, a sophisticated version of the genetic engineering that has been taking place for many millennia on horses, dogs and other animals, which have been, in reality, designed by mankind.

Before we embark, in the next chapters, on a mixture of educated guesses and wild speculations on how the technologies described in the last chapters may one day change the world, we need to better understand how the most complex of organs, the brain, works. In fact, deriving accurate models of the brain is probably the biggest challenge we face in our quest to understand completely the behavior of complex organisms.

8 How the Brain Works

Ever since Alcmaeon of Croton (fifth century BC) and Hippocrates (fourth century BC) recognized the brain as the seat of intelligence, humanity has been interested in understanding how it works. Despite this early start, it took more than two millennia to become common knowledge that it is in the brain that intelligence and memory reside. For many centuries, it was believed that intelligence resided in the heart; even today, we say that something that was memorized was "learned by heart."

The latest estimates put the total number of cells in an average human body at 37 trillion (Bianconi et al. 2013) and the total number of cells in the brain at roughly 86 billion (Azevedo et al. 2009), which means that less than 0.5 percent of our cells are in the brain. However, these cells are at the center of who we are. They define our mind, and when they go they take with them our memories, our personality, and our very essence.

Most of us are strongly attached to our bodies, or to specific physical characteristics of them. Despite the attachment to our bodies, if given the choice most of us probably would prefer to donate a brain than to be the recipient of one, if the transplant of this organ could be performed—something that may actually be attempted in years to come.

Our knowledge of the structures of neurons and other cells in the brain is relatively recent. Although the microscope was invented around the end of the sixteenth century, it took many centuries for microscopes to be successfully applied in the observation of individual brain cells, because individual neurons are more difficult to see than many other types of cells.

Neurons are small, their bodies ranging from a few microns to 100 microns in size. However, they are comparable in size to red blood cells and spermatozoa. What makes neurons difficult to see is the fact that they are intertwined in a dense mass, making it hard to discern individual cells. To borrow an analogy used by Sebastian Seung in his illuminating 2012 book *Connectome*, the neurons in the brain—a mass of intermixed cell bodies, axons, dendritic trees, and other support cells—look like a plate of cooked spaghetti. Even though you can point a microscope at the surface of a section of brain tissue, it isn't possible to discriminate individual cells, because all you see is a tangled mess of cell components.

A significant advancement came in 1873 when Camillo Golgi discovered a method for staining brain tissue that would stain only a small fraction of the neurons, making them visible among the mass of other neurons. We still don't know why the Golgi stain makes only a small percentage of the neurons visible, but it made it possible to distinguish, under a microscope, individual neurons from the mass of biological tissue surrounding them. Santiago Ramón y Cajal (1904) used the Golgi stain and a microscope to establish beyond doubt that nerve cells are independent of one another, to identify many types of neurons, and to map large parts of the brain. Ramón y Cajal published, along with his results, many illustrations of neurons that remain useful to this day. One of them is shown here as figure 8.1.

The identification of neurons and their ramifications represented the first significant advance in our understanding of the brain. Ramón y Cajal put it this way in his 1901 book *Recuerdos de mi vida*: "I expressed the surprise which I experienced upon seeing with my own eyes the wonderful revelatory powers of the chrome-silver reaction and the absence of any excitement in the scientific world aroused by its discovery."

Golgi and Ramón y Cajal shared a Nobel Prize for their contributions, although they never agreed on the way nerve cells interact. Golgi believed that neurons connected with one another, forming a sort of a super-cell; Ramón y Cajal believed that different neurons touched but remained separate and communicated through some yet unknown method. Ramón y Cajal was eventually proved right in the 1920s, when Otto Loewi and Henry Dale demonstrated that neurotransmitters were involved in passing signals between neurons. However, we had to wait until 1954 for the final evidence, when George Palade, George Bennett, and Eduardo de Robertis used the recently invented electron microscope to reveal the structure of synapses. We now know that there are chemical synapses and electrical synapses. Chemical synapses are much more numerous than electrical ones.

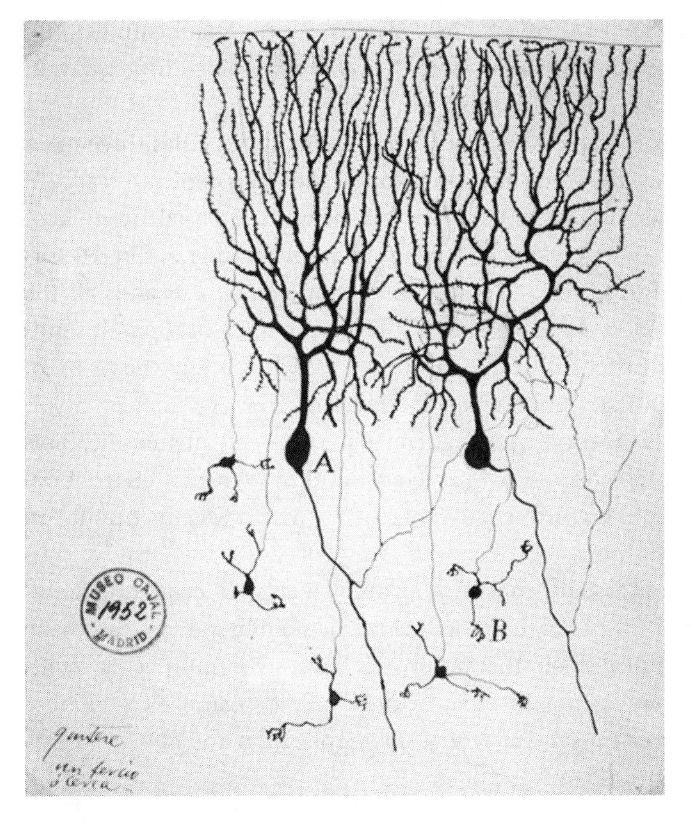

Figure 8.1
A drawing of Purkinje cells (A) and granule cells (B) in the cerebellum of a pigeon by Santiago Ramón y Cajal.

To understand how the brain works, it is important to understand how the individual neurons work, and how they are organized into the amazingly complex system we call the brain.

How Neurons Work

The many different types of neurons can be classified by morphology, by function, or by location. Golgi grouped neurons into two types: those with long axons used to move signals over long distances (type I) and those with short axons (type II). The simplest morphology of type I neurons, of which spinal motor neurons are a good example, consists of a cell body called the *soma* and a long thin axon covered by a myelin sheath. The sheath helps in signal propagation. Branching out from the cell body is a dendritic tree that receives signals from other neurons.

A generic neuron, depicted schematically in figure 8.2a, has three main parts: the dendrites, the soma, and the axon. Figure 8.2b depicts a real neuron from a mouse brain, located in layer 4 of the primary visual area (whose behavior is briefly described later in this chapter). A typical neuron receives inputs in the dendritic tree. Then, in some complex way, it adds all the input contributions in the soma and, if it receives sufficient input, it sends an output through the axon. The output is characterized by the firing of the neuron, which takes place when the neuron receives sufficient input. Neuron firing is a phenomenon generated by the cell membrane. This membrane has specific electrical characteristics that we inherited from the primitive organisms known as choanoflagellates (which were mentioned in chapter 6).

Neurons, like other cells, consist of a complex array of cellular machinery surrounded by a lipid membrane. Inside the membrane, neurons have much of the same machinery that other cells have, swimming inside a saltwater solution: many different types of cytoplasmic organelles (including mitochondria) and a nucleus, in which DNA replication and RNA synthesis

Figure 8.2
(a) A schematic diagram of a neuron by Nicolas Rougier (2007), available at Wikimedia Commons. (b) A real neuron from layer 4 of the primary visual area of a mouse brain, reconstructed and made available by the Allen Institute for Brain Science. Little circles at the end of dendrites mark locations where the reconstruction stopped, while the larger circle marks the location of the soma.

takes place. However, since their job is to carry information from one place to the other in the form of nerve impulses, nerve cell have a different, highly specialized membrane. The membrane of a nerve cell is a complex structure containing many proteins that enable or block the passage of various substances. Of particular interest in the present context are pores, ion channels, and ion pumps of different sizes and shapes. Channels can open or close, and therefore they control the rate at which substances cross the membrane. Most ion channels are permeable only for one type of ion. Ion channels allow ions to move in the direction of the concentration gradient, moving ions from high concentration regions to low concentration regions. Ion pumps are membrane proteins that actively pump ions in or out of the cells, using cellular energy obtained usually from adenosine 5′-triphosphate (ATP) to move the ions against their concentration gradient. Some ion channels—those known as *voltage dependent*—have a pumping capacity that is influenced by the voltage difference across the membrane (known as the *membrane potential*).

The membrane potential, determined by the concentration of charged ions inside and outside the nerve cell, typically ranges from 40 to 80 millivolts (mV), being more positive outside. The cells are bathed in a salt-water solution. The electric potential inside the cell and that outside the cell differ because the concentrations of the ions in the salt water vary across the membrane. Ions flow across the channels, in one direction or the other, while others are pumped by the ion pumps. At rest, there is a voltage difference of roughly 70 mV between the two sides of the membrane. Many ions have different concentrations inside and outside the cell membrane. The potassium ion K^+ has a higher concentration inside than outside; the sodium ion Na^+ and the chloride ion Cl^- have higher concentrations outside than inside. These differences in concentrations of charged ions create the membrane potential.

The existence of the neuron membrane, and the characteristics of the channels and pumps, lead to a behavior of the neurons that can be well modeled by an electrical circuit. With these general concepts in mind, we can now try to understand how a detailed electrical model of a neuron can be used to simulate its behavior. Let us first consider an electrical model of a section of the membrane.

The important part of the behavior of a neuron for the purposes of signal transmission consists in the opening and the closing of channels and pumps that transport the ions across the membrane. It is possible to derive a detailed electrical model for a small section of the membrane, and to interconnect these sections to obtain a working model for the whole neuron. The model for a segment of passive membrane can therefore be a simple electrical circuit, such as that illustrated in figure 8.3. Attentive readers will notice many resemblances between this circuit and the circuit that was used in figure 3.2 to illustrate Maxwell's equations.

In figure 8.3 the voltage sources V_K and V_{Na} stand for the equilibrium potential created by the differences in ion concentrations (only K^+ and Na^+ are considered in this example). The conductances G_K and G_{Na} represent the combined effect of all open channels permeable to ions. The capacitor models the membrane capacitance. It can be assumed, without loss of modeling power, that the outside of the neuron is at zero potential. The passive membrane of a neuron is therefore represented by a large number of circuits like this one, each corresponding to one patch of the membrane, interconnected by conductances in the direction of signal transmission. The two voltage sources that model the sodium and potassium ion concentrations can be replaced by an electrically equivalent voltage source and conductance, called the Thévenin equivalent; the resulting circuit is illustrated in figure 8.4.

Alan Hodgkin and Andrew Huxley have extensively studied the electrical model of the neuron membrane in the presence of the many different types of ion pumps and channels that are present in real neurons. The behavior of an active membrane is much more complex than the passive model shown, the added complexity being due to the time-dependent and voltage-dependent variation of the conductances.

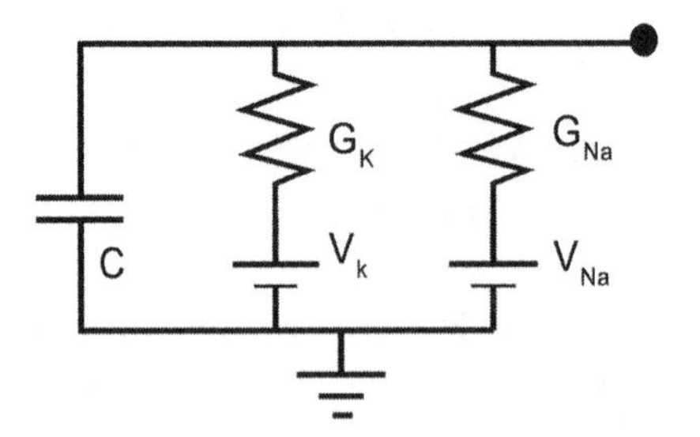

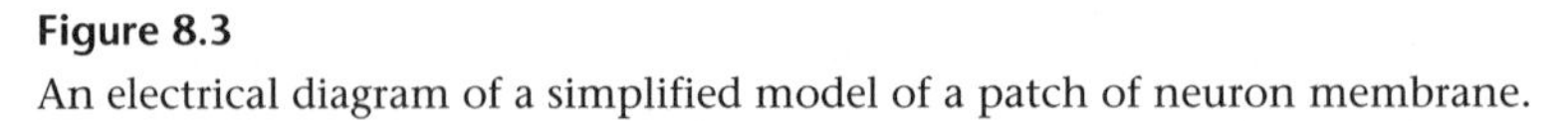
Figure 8.3
An electrical diagram of a simplified model of a patch of neuron membrane.

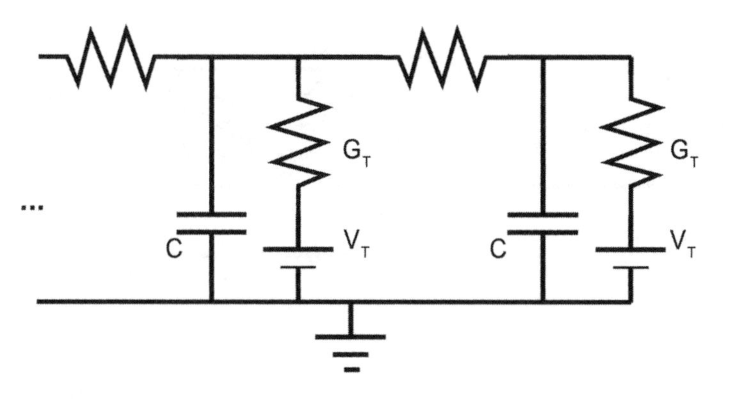

Figure 8.4
Simplified models of patches of neuron membrane interconnected by axial conductances.

However, extensive studies conducted by Hodgkin, Huxley, and many other researchers that followed them have told us how neuron membranes work in great detail.

In their experiments, Hodgkin and Huxley used the giant axon of the squid to obtain and validate models for ionic mechanisms of action potentials. That axon, which is up to a millimeter in diameter, controls part of the squid's water-jet propulsion system. Its size makes it amenable to experiments and measurements that would otherwise be very difficult. The Hodgkin-Huxley model describes in detail the behavior of time-dependent and voltage-dependent conductances that obey specific equations. The model for an active segment of the membrane then becomes a generalization of the one shown in figure 8.4, with conductances that vary with time and voltage. For each conductance, there is a specific equation that specifies its value as a function of time and the voltage across it. With this model, Hodgkin and Huxley (1952) were able to simulate the generation of axon potentials in the squid giant axon and to accurately predict effects not explicitly included in the models. In particular, they predicted the axonal propagation speed with a good accuracy, using parameters obtained experimentally. Different generalizations of this model for particular types of neurons were obtained through the work of many other researchers.

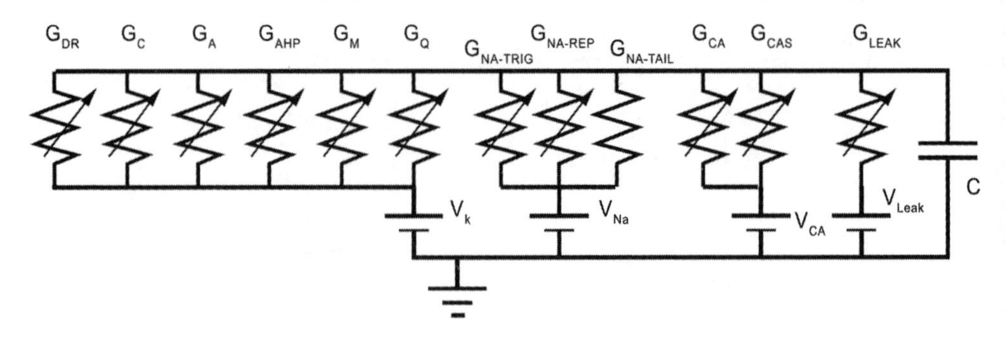

Figure 8.5
A complete electrical diagram of Hodgkin and Huxley's model of a patch of neuron membrane of a pyramidal neuron.

Figure 8.5 illustrates the components of the model for an active segment of a membrane from a pyramidal neuron in the human brain. Each conductance shown corresponds to a particular ion conductance mechanism in the membrane. If we plug in the detailed time and voltage dependences of each of the parameters involved in this circuit, and interconnect many of these circuits, one for each patch of the membrane, we obtain an electrical model for a complete neuron.

When a neuron isn't receiving input from other neurons, the value of the membrane potential is stable and the neuron doesn't fire. Firing occurs only when the neuron receives input from other neurons through the synapses that interconnect the neurons, usually connecting the axon of the pre-synaptic neuron to a dendrite in the post-synaptic neuron. When enough neurons with their axons connected to the dendritic tree of this neuron are firing, electrical activity in the pre-synaptic neurons is converted, by the activity of the synapses, into an electrical response that results in an increase (hyperpolarization) or a decrease (depolarization) of the membrane potential in the receiving neuron.

If one section of the membrane in the receiving neuron—most commonly a section at the base of the axon (Stuart et al. 1997)—is sufficiently depolarized, one observes, both in computer simulations and in real neurons, that the membrane depolarizes further, which leads to a rapid self-sustained decrease of the membrane potential (called an *action potential*).

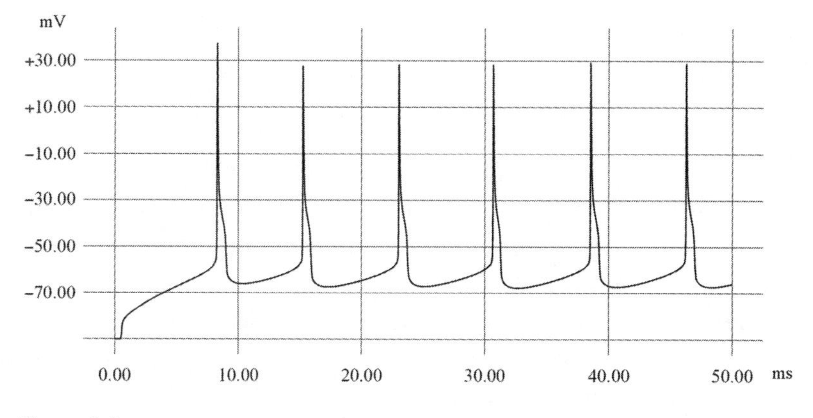

Figure 8.6
A simulation of the electrical behavior of a pyramidal neuron.

This rapid self-sustained decrease of the membrane potential corresponds to the well-known spiking behavior of neurons. When a neuron spikes, the membrane potential depolarizes suddenly, recovering a value closer to its resting state in a few milliseconds. Figure 8.6 shows the resulting voltages across the membrane (obtained using a simulator).

When the membrane voltage reaches a certain level, it undergoes a regenerative process that generates a spike. After firing, a neuron has a refractory period, during which it isn't able to generate other spikes even if excited. The length of this period varies widely from neuron to neuron. The spiking depicted in figure 8.6 corresponds to the evolution in time of the voltage in a section of the neuron membrane. However, since this section of the membrane is connected to adjacent sections, this depolarization leads to depolarization in the adjacent sections, which causes the spike to propagate through the axon until it reaches the synaptic connections in the axon terminals. The characteristics of active membranes are such that, once an impulse is initiated, it propagates at a speed that depends only on the geometry of the axon. In fact, once the impulse is initiated, its effects are essentially independent of the waveform in the soma or in the dendritic tree where it started. It is, in a way, a digital signal, either present or absent. However, the frequency and perhaps the timing of the spikes is used to encode information transmitted between neurons.

When the spike reaches a chemical synapse that makes a connection with another neuron, it forces the release of neurotransmitter molecules from the synapse vesicles in the pre-synaptic membrane. These molecules attach to receivers in the post-synaptic membrane and change the state of ionic channels in that membrane. These changes create a variation in the ion fluxes that depolarizes (in excitatory connections) or hyperpolarizes (in inhibitory connections) the membrane of the receiving neuron. Those changes in the voltage drop across the membrane are known, respectively, as *excitatory post-synaptic potentials* (EPSP) and *inhibitory post-synaptic potentials* (IPSP). Chemical synapses, the most common synapses, are located in gaps between the membranes of the pre-synaptic and post-synaptic neurons that range from 20 to 40 nanometers.

The brain also contains electrical synapses. An electrical synapse contains channels that cross the membranes of the target neurons and allow ions to flow from the synapse to the next neuron, thereby transmitting the signal. When the membrane potential of the pre-synaptic neuron changes, ions may move through these channels and transmit the signal. Electrical synapses conduct nerve impulses faster than chemical synapses, but they do not provide electrical gain. For that reason, the signal in the post-synaptic neuron is an attenuated version of the signal in the originating neuron.

Our current knowledge of the detailed workings of the neuron membrane enables us to simulate, with great accuracy, the behavior of single neurons or of networks of neurons. Such simulation requires detailed information about the structure and electrical properties of each neuron and about how the neurons are interconnected, including the specific characteristics of each synapse. A number of projects (including some described in the next chapter, such as the Allen Brain Atlas) aim at developing accurate models for neurons in complex brains. This is done by choosing specific neurons and measuring, in great detail, their electrical response to a number of different stimuli. These stimuli are injected into the neuron using very thin electrical probes. The responses obtained can then be used to create and tune very precise electrical models of neurons.

Neurons of more complex organisms are much smaller and more diverse than the neurons Hodgkin and Huxley studied, and are interconnected in a very complex network with many billions of neurons and many trillions of synapses. Understanding the detailed organization of this complex network (perhaps the most complex task ever undertaken) could lead to fundamental changes in medicine, in technology, and in society. This objective is being pursued in many, many ways.

The Brain's Structure and Organization

A modern human brain has nearly 100 billion neurons, each of them making connections, through synapses, with many other neurons. The total number of synapses in a human brain is estimated to be between 10^{14} and 10^{15}, which gives an average number of synapses per neuron between 1,000 and 10,000. Some neurons, however, have much more than 10,000 synapses.

Neuroanatomy has enabled us to identify the general functions and characteristics of many different areas of the brain. The different characteristics of gray matter (composed mainly of neuron bodies) and white matter (composed mainly of neuron axons) have been known for centuries. Some parts of the brain (including the cortex) are composed mostly of gray matter; others (including the corpus callosum, a structure that interconnects the two hemispheres) are composed mostly of white matter. However, exactly how the brain's various areas work remains largely a mystery, although some areas are better understood than others. The brain is usually considered to be divided into three main parts: the forebrain, the midbrain, and the hindbrain, each of them subdivided in a number of areas. Each area has been associated with a number of functions involved in the behavior of the body.

The cerebrum (a part of the forebrain) is the largest part of the human brain, and is commonly associated with higher brain functions, including memory, problem solving, thinking, and feeling. It also controls movement. In general, the closer an area is to the sensory inputs, the better it is understood. The cortex, the largest part of the cerebrum, is of special interest, because it is involved in higher reasoning and in the functions we associate with cognition and intelligence. It is believed to be more flexible and adjustable than the more primitive parts of the brain. The cortex is a layer of neural tissue, between 2 and 4 millimeters thick, that covers most of the brain. The cortex is folded in order to increase the amount of cortex surface area that can fit into the volume available within the skull. The pattern of folds is similar in different individuals but shows many small variations.

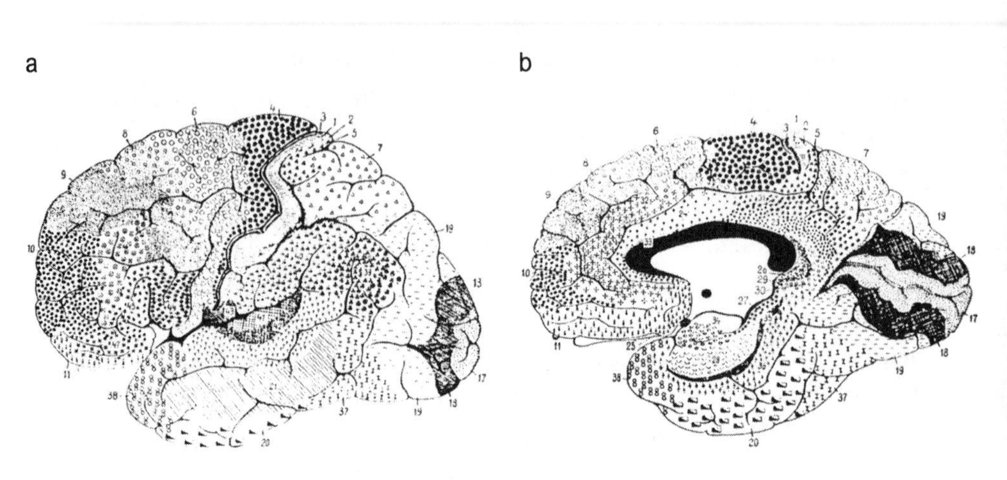

Figure 8.7
A diagram of the Brodmann areas reprinted from Ranson and Saunders 1920. (a) Lateral surface. (b) Medial surface.

The anatomist Korbinian Brodmann defined and numbered brain cortex areas mostly on the basis of the cellular composition of the tissues observed with a microscope. (See Brodmann 1909.) On the basis of his systematic analysis of the microscopic features of the cortex of humans and several other species, Brodmann mapped the cortex into 52 areas. (See figure 8.7.) The results Brodmann published in 1909 remain in use today as a general map of the cortex.

Brodmann's map of the human cortex remains the most widely known and frequently cited, although many other studies have proposed alternative and mode detailed maps. Brodmann's map has been discussed, debated, and refined for more than a hundred years. Many of the 52 areas Brodmann defined on the basis of their neuronal organization have since been found to be closely related to various cortical functions. For example, areas 1–3 are the primary somatosensory cortex, areas 41 and 42 correspond closely to primary auditory cortex, and area 17 is the primary visual cortex. Some Brodmann areas exist only in non-human primates. The terminology of Brodmann areas has been used extensively in studies of the brain employing many different technologies, including electrode implantation and various imaging methods.

Studies based on imaging technologies have shown that different areas of the brain become active when the brain executes specific tasks. Detailed "atlases" of the brain based on the results of these studies (Mazziotta et al. 2001; Heckemann et al. 2006) can be used to understand how functions are distributed in the brain. In general, a particular area may be active when the brain executes a number of different tasks, including sensory processing, language usage or muscle control.

A particularly well-researched area is the visual cortex. Area 17 has been studied so extensively that it provides us with a good illustration of how the brain works. The primary visual area (V1) of the cerebral cortex, which in primates coincides with Brodmann area 17, performs the first stage of cortical processing of visual information. It is involved in early visual processing, in the detection of patterns, in the perception of contours, in the tracking of motion, and in many other functions. Extensive research performed in this area has given us a reasonably good understanding of the way it operates: It processes information received from the retina and transforms the information into high-level features, such as edges, contours and line movement. Its output is then fed to upstream visual areas V2 and V3.

The detailed workings of the retina (an extremely complex system in itself) have been studied extensively, and the flow of signals from the retina to the visual cortex is reasonably well understood. In the retina, receptors (cones and rods) detect incoming photons and perform signal processing, the main purpose of which is to detect center-surround features (a dark center in a light surround, or the opposite). Retinal ganglion cells, sensitive to these center-surround features, send nervous impulses through the optic nerve into the *lateral geniculate nucleus* (LGN), a small, ovoid part of the brain that is, in effect, a relay center. The LGN performs some transformations and does some signal processing on these inputs to obtain three-dimensional information. The LGN then sends the processed signals to the visual cortex and perhaps to other cortical areas, as illustrated in figure 8.8.

Electrode recording from the cortex of living mammals, pioneered by David Hubel and Torsten Wiesel (1962, 1968), has enabled us to understand how cells in the primary visual cortex process the information coming from the lateral geniculate nucleus. Studies have shown that the primary visual cortex consists mainly of cells responsive to simple and complex features in the input.

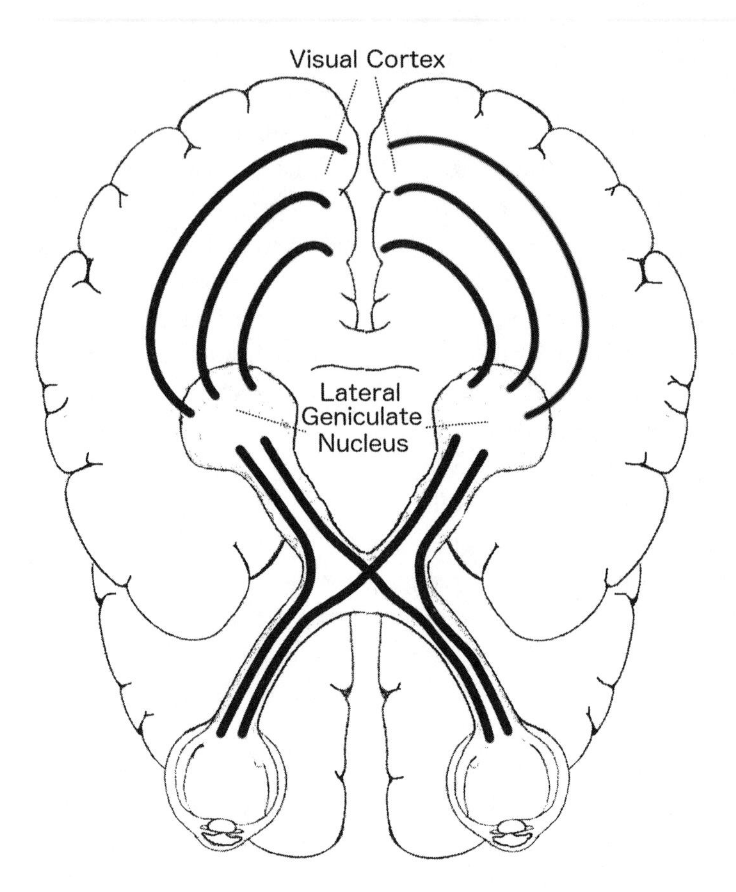

Figure 8.8
Neural pathways involved in the first phases of image processing by the brain.

The groundbreaking work of Hubel and Wiesel, described beautifully in Hubel's 1988 book *Eye, Brain, and Vision,* advanced our understanding of the way we perceive the world by extracting relevant features from the images obtained by the retina. These features get more and more complex as the signals go deeper and deeper into the visual system. For example, *ocular dominance columns*—groups of cells in the visual cortex—respond to visual stimuli received from one eye or the other. Ocular dominance columns are groups of neurons, organized in stripes in the visual cortex, that respond preferentially to input from either the left eye or the right eye. The columns, laid out in striped patterns across the surface of the primary visual cortex, span multiple cortical layers and detect different features. The particular features detected vary across the surface of the cortex, continuously, in complex patterns called *orientation columns*. Simple cells, as they are known, detect the presence of a line in a particular part of the retina—either a dark line surrounded by lighter areas or the opposite, a light line surrounded by darker areas. Complex cells perform the next steps in the analysis. They respond to a properly oriented line that sweeps across the receptive field, unlike simple cells that respond only to a stationary line critically positioned in one particular area of their receptive field. Complex cells in the cortex exhibiting many different functions have been found. Some respond more strongly when longer edges move across their receptive field, others are sensitive to line endings, and others are sensitive to different combinations of signals coming from simple cells.

The architecture of the primary visual cortex derived from the experiments mentioned above and from other experiments has given us a reasonably good understanding of the way signals flow in this area of the brain. Incoming neurons from the lateral geniculate nucleus enter mainly in layer 4 of the cortex, relaying information to cells with center-surround receptive fields and to simple cells. Layers 2, 3, 5, and 6 consist mainly of complex cells that receive signals from simple cells in the different sub-layers of layer 4.

The workings of other areas are not yet understood as well as the workings of area V1. Despite the extensive brain research that has taken place in recent decades, only fairly general knowledge of the roles of most of the brain's areas and of how they operate has been obtained so far. Among the impediments to a more detailed understanding are the complexity of the areas further upstream in the signal-processing pipeline, the lack of a principled approach to explaining how the brain works, and the limitations of the mechanisms currently available to obtain information about the detailed behavior of brain cells.

We know, however, that it is not correct to view the brain as a system that simply processes sensory inputs and converts them into actions (Sporns 2011). Even when not processing inputs, the brain exhibits spontaneous activity, generating what are usually called *brain waves*. These waves, which occur even during resting states, have different names, depending on the frequency of the electromagnetic signals they generate: alpha (frequencies in the range 8–13 Hz), beta (13–35 Hz), gamma (35–100 Hz), theta (3–8 Hz), and delta (0.5–3 Hz). The ranges are indicative and not uniquely defined, but different brain waves have been associated with different types of brain states, such as deep sleep, meditation, and conscious thought. But despite extensive research on the roles of spontaneous neural activity and of the resulting brain waves, we still know very little about the roles this activity plays (Raichle 2009). What we know is that complex patterns of neural activity are constantly active in the brain, even during deep sleep, and that they play important but mostly unknown roles in cognition. These patterns of activity result from the oscillations of large groups of neurons, interconnected into complex feedback loops, at many different scales, which range from neighboring neuron connections to long-range interconnections between distant areas of the brain.

A complete understanding of the function and the behavior of each part of the brain and of each group of neurons will probably remain outside the range of the possible for many years. Each individual brain is different, and it is likely that only general rules about the brain's organization and its functioning will be common to different brains. At present, trying to understand the general mechanisms brains use to organize themselves is an important objective for brain sciences, even in the absence of a better understanding of the detailed functions of different areas of the brain.

Brain Development

When we speak of understanding the brain, we must keep in mind that a complete understanding of the detailed workings of a particular brain will probably never be within the reach of a single human mind. In the same way, no human mind can have a complete understanding of the detailed workings of a present-day computer, bit by bit. However, as our knowledge advances, it may become possible to obtain a clear understanding of the general mechanisms used by brains to organize themselves and to become what they are, in the same way that we have a general understanding of the architecture and mechanisms used by computers.

This parallel between understanding the brain and understanding a computer is useful. Even though no one can keep in mind the detailed behavior of a present-day computer, humans have designed computers, and therefore it is fair to say that humans understand computers. Human understanding of computers doesn't correspond to detailed knowledge, on the part of any individual or any group, of the voltage and current in each single transistor. However, humans understand the general principles used to design a computer, the way each part is interconnected with other parts, the behavior of each different part, and how these parts, working together, perform the tasks they were designed to perform. With the brain, things are much more complex. Even though there are general principles governing the brain's organization, most of them are not yet well understood. However, we know enough to believe that the brain's structure and organization result from a combination of genetic encoding and brain plasticity.

It is obvious that genetic encoding specifies how the human brain is constructed of cells built from proteins, metabolites, water, and other constituents. Many genes in the human genome encode various properties of the cells in the human brain and (very, very indirectly) the structure of the human brain. However, the genetic information doesn't specify the details of each particular neuron and each particular connection between neurons. The genetic makeup of each organism controls the way brain cells are created and duplicated, and controls the large-scale architecture of the brain, as well as the architecture of the other components of the body. However, in humans and other higher organisms the genetic makeup doesn't control the specific connections neurons make with one another, nor does it control the activity patterns that characterize the brain's operation.

Extensive studies on brain development in model organisms and in humans are aimed at increasing our understanding of the cellular and molecular mechanisms that control the way nervous systems are created during embryonic development. Researchers working in developmental biology have used different model organisms, including the ones referred to in chapter 7, to study how brains develop and self-organize. They have found that, through chemical gradients and other mechanisms, genetic information directs the neurons to grow their axons from one area of the brain to other areas. However, the detailed pattern of connections that is established between neurons is too complex to be encoded uniquely by the genes, and too dynamic to be statically defined by genetic factors. What is encoded in the genome is a set of recipes for how to make neurons, how to control the division of neuron cells and the subsequent multiplication of neurons, how to direct the extension of the neuron axons and dendritic trees, and how to control the establishment of connections with other neurons.

Current research aims at understanding the developmental processes that control, among other things, the creation and differentiation of neurons, the migration of new neurons from their birthplace to their final positions, the growth and guidance of axon cones, the creation of synapses between these axons and their post-synaptic partners, and the pruning of many branches that takes place later. Many of these processes are controlled in a general way by the genetic makeup of the organism and are independent of the specific activities of particular neuron cells.

The formation of the human brain begins with the neural tube. It forms, during the third week of gestation, from the neural progenitor cells located in a structure called the *neural plate* (Stiles and Jernigan 2010). By the end of the eighth week of gestation, a number of brain regions have been formed and different patterns of brain tissue begin to appear.

The billions of neurons that constitute a human brain result from the reproduction of neural progenitor cells, which divide to form new cells. Neurons are mature cells and do not divide further to give origin to other neurons. Neural progenitor cells, however, can divide many times to generate two identical neural progenitor cells capable of further division. These cells then produce either neurons or glial cells, according to the biochemical and genetic regulation signals they receive. Glial cells (also called *glia* or *neuroglia*) are non-neuronal cells that maintain the chemical equilibrium in the brain and provide physical support for neurons. They are also involved in the generation of myelin (an insulator that surrounds some axons, increasing their transmission speed and avoiding signal loss).

In humans, the generation of new neurons in the cortex is complete around the sixteenth week of gestation (Clancy, Darlington, and Finlay 2001). After they are generated, neurons differentiate into different types. Neuron production is located mainly in an area that will later become the ventricular zone. The neurons produced in that region migrate into the developing neocortex and into other areas. The various mechanisms used by neurons to migrate have been studied extensively, but the process is extremely complex and is only partially understood (Kasthuri et al. 2015; Edmondson and Hatten 1987).

Once a neuron has reached its target region, it develops axons and dendrites in order to establish connections with other neurons. Neurons create dense arbors of dendrites in their immediate vicinity, and extend their axons by extending the axons' growth cones, guided by chemical gradients that direct the growth cones toward their intended targets. The fact that some of the molecules used to guide the growth of axons are attractive and others are repulsive results in a complex set of orientation clues perceived by the growth cones. Once the axons reach their target zone, they establish synapses with dendritic trees in the area.

A significant fraction of the neurons that develop during this process and a significant fraction of the connections they establish disappear in the next stages (pre-natal and post-natal) of the brain's development. A significant fraction of the neurons die. Even in those that don't die, many connections are pruned and removed. Initial connection patterns in the developing brain involve many more synapses than the ones that remain after the brain reaches its more stable state, in late childhood. Overall, the total number of established synapses may be cut by half relatively to the peak value it reached in early childhood.

The processes of neuron migration, growth-cone extension, and pruning are controlled, in large part, by genetic and biochemical factors. However, this information isn't sufficient to encode the patterns in a fully formed brain. Brain plasticity also plays a very significant role. The detailed pattern of connections in a fully formed brain results in large part from activity-dependent mechanisms in which the detailed activity patterns of the neurons, resulting from sensory experience and from spontaneous neuron activation, control the formation of new synapses and the pruning of existing ones.

Plasticity, Learning, and Memory

The human brain is plastic (that is, able to change) not only during its development but throughout a person's life. Plasticity is what gives a normal brain the ability to learn and to modify its behavior as new experiences occur. Though plasticity is strongest during childhood, it remains a fundamental and significant property of the brain throughout a person's life.

The brain's plasticity comes into play every time we see or hear something new, every time we make a new memory, and every time we think. Even though we don't have a complete understanding of the processes that create memories, there is significant evidence that long-term memories are stored in the connection patterns of the brain. Short-term memories are likely to be related to specific patterns of activity in the brain, but those patterns are also related to changes (perhaps short-lived changes) in connectivity patterns, which means that brain plasticity is active every second of our lives.

The connections between the neurons are dynamic, and they change as a consequence of synaptic plasticity. Synaptic plasticity is responsible not only for the refinement of newly established neural circuits in early infancy, but also for the imprinting of memories later in life and for almost all the mechanisms that are related to learning and adaptability. The general principles and rules that control the plasticity of synapses (and, therefore, of the brain) are only partially understood.

Santiago Ramón y Cajal was probably the first to suggest that there was a learning mechanism that didn't require the creation of new neurons. In his 1894 Croonian Lecture to the Royal Society, he proposed that memories might be formed by changing the strengths of the connections between existing neurons.

Donald Hebb, in 1949, followed up on Ramón y Cajal's ideas and proposed that neurons might grow new synapses or undergo metabolic changes that enhance their ability to exchange information. He proposed two simple principles, which have hence been found to be present in many cases. The first states that the repeated and simultaneous activation of two neurons leads to a reinforcement of connections between them. The second states that, if two neurons are repeatedly active sequentially, then the connections from the first to the second become strengthened (Hebb 1949). This reinforcement of the connections between two neurons that fire in a correlated way came to be known as Hebb's Rule. Hebb's original proposal was presented as follows in his 1949 book *The Organization of Behavior*:

> Let us assume that the persistence or repetition of a reverberatory activity (or "trace") tends to induce lasting cellular changes that add to its stability. … When an axon of cell A is near enough to excite a cell B and repeatedly or persistently takes part in firing it, some growth process or metabolic change takes place in one or both cells such that A's efficiency, as one of the cells firing B, is increased.

Today these principles are often rephrased to mean that changes in the efficacy of synaptic transmission result from correlations in the firing activity of pre-synaptic and post-synaptic neurons, leading to the well-known statement "Neurons that fire together wire together." The fact that this formulation is more general than Hebb's original rule implies that a neuron that contributes to the firing of another neuron has to be active slightly before the other neuron. This idea of correlation-based learning, now generally called *Hebbian learning*, probably plays a significant role in the plasticity of synapses.

Hubel and Wiesel studied the self-organization of the visual cortex by performing experiments with cats and other mammals that were deprived of vision in one eye before the circuits in the visual cortex had had time to develop (Hubel and Wiesel 1962, 1968; Hubel 1988). In cats deprived of the use of one eye, the columns in the primary visual cortex rearranged themselves to take over the areas that normally would have received input from the deprived eye. Their results showed that the development of cortical structures that process images (e.g., simple cells and complex cells) depends on visual input. Other experiments performed with more specific forms of visual deprivation confirmed those findings. In one such experiment, raising cats from birth with one eye able to view only horizontal lines and the other eye able to view only vertical lines led to a corresponding arrangement of the ocular dominance columns in the visual cortex (Hirsch and Spinelli 1970; Blakemore and Cooper 1970). The receptive fields of cells in the visual cortex were oriented horizontally or vertically depending on which eye they were sensitive to, and no cells sensitive to oblique lines were found. Thus, there is conclusive evidence that the complex arrangements found in the visual cortex are attributable in large part to activity-dependent plasticity that, in the case of the aforementioned experiment, is active only during early infancy. Many other results concerning the visual cortex and areas of the cortex dedicated to other senses have confirmed Hubel and Wiesel's discoveries. However, the mechanisms that underlie this plasticity are still poorly understood and remain subjects of research.

A number of different phenomena are believed to support synaptic plasticity. Long-term potentiation (LTP), the most prominent of those phenomena, has been studied extensively and is closely related to Hebb's Rule. The term LTP refers to a long-term increase in the strength of synapses in response to specific patterns of activity that involved the pre-synaptic and post-synaptic neurons. The opposite of LTP is called *long-term depression* (LTD). LTP, discovered in the rabbit hippocampus by Terje Lømo (1966), is believed to be among the cellular mechanisms that underlie learning and memory (Bliss and Collingridge 1993; Bliss and Lømo 1973).

Long-term potentiation occurs in a number of brain tissues when the adequate stimuli are present, but it has been most studied in the hippocampus of many mammals, including humans (Cooke and Bliss 2006). LTP is expressed as a persistent increase in the neural response in a pathway when neural stimuli with the right properties and appropriate strength and duration are present. It has been shown that the existence of LTP and the creation of memories are correlated, and that chemical changes which block LTP also block the creation of memories. This result provides convincing evidence that long-term potentiation is at least one of the mechanisms, if not the most important one, involved in the implantation of long-term memories.

Another phenomenon that may play a significant role in synaptic plasticity is called *neural back-propagation*. Despite its name, the phenomenon is only vaguely related to the back-propagation algorithm that was mentioned in chapter 5. In neural back-propagation, the action potential that (in most cases) originates at the base of the axon also creates an action potential that goes back, although with decreased intensity, into the dendritic arbor (Stuart et al. 1997). Some researchers believe that this simple process can be used in a manner similar to the back-propagation algorithm, used in multi-layer perceptrons, to back-propagate an error signal; however, not enough evidence of that mechanism has been uncovered so far.

Synaptic plasticity is not the only mechanism underlying brain plasticity. Although creation, reinforcement, and destruction of synapses are probably the most important mechanisms that lead to brain plasticity, other mechanisms are also likely to be involved. Adult nerve cells do not reproduce, and therefore no new neurons are created in adults (except in the hippocampus, where some stem cells can divide to generate new neurons). Since, in general, there is no creation of new nerve cells, brain plasticity isn't likely to happen through the most obvious mechanism, the creation of new neurons. However, mechanisms whereby existing neurons migrate or grow new extensions (axons or dendritic trees) have been discovered, and they may account for a significant amount of adults' brain plasticity.

Further research on the mechanisms involved in brain plasticity and on the mechanisms involved in the brain's development will eventually lead to a much clearer understanding of how the brain organizes itself. Understanding the principles behind the brain's development implies the creation and simulation of much better models for the activity-independent mechanisms that genes use to control the brain's formation and the activity-dependent mechanisms that give it plasticity.

To get a better grasp of the current status of brain science, it is interesting to look at the projects under way in this area and at the technologies used to look deep inside the brain. Those are the topics of the next chapter.

9 Understanding the Brain

For many reasons, understanding the mechanisms that underlie the development and the plasticity phenomena of the human brain would be of enormous value. From a medical standpoint, the understanding of these mechanisms would lead to ways to prevent degenerative brain disorders, to extend life, and to increase the length of time people live in full possession of their mental abilities. By themselves, those objectives would justify the enormous effort now under way to understand in detail how the brain operates. However, such an understanding would also be valuable from an engineering standpoint, because deep knowledge of the mechanisms by which the brain operates would also give us the tools necessary to recreate the same behaviors in a computer.

To advance our understanding of the brain, we need to obtain extensive and detailed information about its internal organization and about how its activation patterns encode thoughts. Advances in instrumentation technology have given us many ways to observe a living brain. The objective is to improve our knowledge of the structures and mechanisms involved in the creation and operation of a brain without disturbing its behavior. Techniques that involve the destruction of the brain can provide much finer complementary information about neuron structure and connectivity.

A number of very large research projects and a multitude of smaller ones aim at obtaining information about brain structure and behavior that can be used to reveal how brains work, but also to understand a number of other important things.

In the United States, the multi-million-dollar Human Connectome Project (HCP) aims to provide an extensive compilation of neural data and a graphic user interface for navigating through the data. The objective is to obtain new knowledge about the living human brain by building a "network map" that will provide information on the anatomical and functional connectivity within a human brain and to use this knowledge to advance our understanding of neurological disorders. Another major project, the Open Connectome Project, aims at creating publicly available connectome data from many organisms.

Also in the United States, the Brain Research through Advancing Innovative Neurotechnologies (BRAIN) initiative, also known as the Brain Activity Map Project, is a multi-billion-dollar research initiative announced in 2013, with the goal of mapping the activity of every neuron in the human brain. The idea is that, by accelerating the development and application of new technologies, it will be possible to obtain a new dynamic view of the brain that will show, for the first time, how individual cells and complex neural circuits interact.

In Europe, the Human Brain Project (HBP) is a ten-year project, with a budget of more than a billion euros, financed largely by the European Union. Established in 2013, it is coordinated by the École Polytechnique Fédérale de Lausanne, the same university that coordinated the Blue Brain Project, a project that obtained some of the most significant research results to date in brain simulation. The Human Brain Project aims to achieve a unified, multi-level, understanding of the human brain by integrating data about healthy and diseased brains. The project focuses on the data that will have to be acquired, stored, organized, and mined in order to identify relevant features in the brain. One of its main objectives is the development of novel neuromorphic and neuro-robotic technologies based on the brain's circuitry and computing principles.

In Japan, Brain/MINDS (Brain Mapping by Integrated Neurotechnologies for Disease Studies) is a multi-million-dollar project, launched in 2014, that focuses on using non-human primate brains to obtain a better understanding of the workings of the human brain. As is also true of the other projects, one of the principal aims of Brain/MINDS is to elucidate the mechanisms involved in brain disorders. The idea is to use the marmoset, a small primate, as a model to study cognition and neural mechanisms that lead to brain disorders. (Simpler model organisms, such as the mouse, may be too far away evolutionarily from humans to provide adequate platforms to study the human brain. Using marmosets may help circumvent this limitation.)

The Allen Institute for Brain Science is a nonprofit private research organization that conducts a number of projects aimed at understanding how the human brain works. The Allen Human Brain Atlas (Hawrylycz et al. 2012), now under development, is a highly comprehensive information system that integrates data collected by means of live brain imaging, tissue microscopy, and DNA sequencing to document many different pieces of information about the brain of mice, non-humans primates, and humans. The information made available includes where in the brain certain genes are active, brain connectivity data, and data about the morphology and behavior of specific neuron cells.

Looking Inside

The above-mentioned projects, and many other projects that address similar matters, use various technologies to derive detailed information about the brains of humans, primates, and other model organisms. Common to many of these projects are the techniques used to obtain information about brain structures and even, in some cases, about neuron-level connectivity.

In this section we will consider imaging methods that can be used, in non-invasive ways, to obtain information about working brains in order to observe their behavior. Such methods are known as *neuroimaging*. When the objective is to obtain detailed information about the three-dimensional structure and composition of brain tissues, it is useful to view the brain as divided into a large number of voxels. A voxel is the three-dimensional equivalent of a pixel in a two-dimensional image. More precisely, a voxel is a three-dimensional rectangular cuboid that corresponds to the fundamental element of a volume, with the cuboid dimensions imposed by the imaging technology. Cuboid edge sizes range from just a few nanometers to a centimeter or more, depending on the technology and the application. Smaller voxels contain fewer neurons on average, and correspond to lower levels of neuronal activity. Therefore, the smaller the voxels, the harder it is to obtain accurate information about their characteristics using imaging techniques that look at the levels of electrical or chemical activity. Smaller voxels also take longer to scan, since scanning time, in many technologies, increases with the number of voxels. With existing technologies, a voxel used in the imaging of live brains will typically contain a few million neurons and a few billion synapses, the actual number depending on the voxel size and the region of the brain being imaged. Voxels are usually arranged in planes, or slices, which are juxtaposed to obtain complete three-dimensional information about the brain. Many of the techniques used to image the brain are also used to image other parts of the body, although some are specifically tuned to the particular characteristics of brain tissue.

Neuroimaging uses many different physical principles to obtain detailed information about the structure and behavior of the brain. Neuroimaging can be broken into two large classes: *structural imaging* (which obtains information about the structure of the brain, including information about diseases that manifest themselves by altering the structure) and *functional imaging* (which obtains information about brain function, including information that can be used to diagnose diseases affecting function, perhaps without affecting large structures).

Among the technologies that have been used in neuroimaging (Crosson et al. 2010) are computed tomography (CT), near-infrared spectroscopy (NIRS), positron-emission tomography (PET), a number of variants of magnetic-resonance imaging (MRI), electroencephalography (EEG), magnetoencephalography (MEG), and event-related optical signal (EROS).

Computed tomography (Hounsfield 1973), first used by Godfrey Hounsfield in 1971 at Atkinson Morley's Hospital in London, is an imaging technique that uses computer-processed x-ray images to produce tomographic images which are virtual slices of tissues that, stacked on top of each other, compose a three-dimensional image. X rays with frequencies between 30 petahertz (3×10^{16} Hz) and 30 exahertz (3×10^{19} Hz) are widely used to image the insides of objects, since they penetrate deeply in animal tissues but are attenuated to different degrees by different materials.

Tomographic images enable researchers to see inside the brain (and other tissues) without cutting. Computer algorithms process the received x-ray images and generate a three-dimensional image of the inside of the organ from a series of two-dimensional images obtained by sensors placed outside the body. Usually the images are taken around a single axis of rotation; hence the term *computed axial tomography* (CAT). Computed tomography data can be manipulated by a computer in order to highlight the different degrees of attenuation of an x-ray beam caused by various body tissues. CT scans may be done with or without the use of a *contrast agent* (a substance that, when injected into the organism, causes a particular organ or tissue to be seen more clearly with x rays). The use of contrast dye in CT angiography gives good visualization of the vascular structures in the blood vessels in the brain.

Whereas x-ray radiographs have resolutions comparable to those of standard photographs, computed tomography only reaches a spatial resolution on the order of a fraction of a millimeter (Hsieh 2009). A CT scan takes only a few seconds but exposes the subject to potentially damaging ionizing radiation. Therefore, CT scans are not commonly used to study brain behavior, although they have provided important information about brain macro-structures. CT scans are used mostly to determine changes in brain structures that occur independently of the level of activity.

Because the characteristics of the tissue change slightly when the neurons are firing, it is possible to visualize brain activity using imaging techniques. One of the most significant effects is the change in blood flow. When a specific area of the brain is activated, the blood volume in the area changes quickly. A number of imaging techniques, including NIRS, PET, and MRI, use changes in blood volume to detect brain activity. The fact that water, oxygenated hemoglobin, and deoxygenated hemoglobin absorb visible and near-infrared light can also be used to obtain information about the location of neuronal activity.

Blood oxygenation varies with levels of neural activity because taking neurons back to their original polarized state after they have become active and fired requires pumping ions back and forth across the neuronal cell membranes, which consumes chemical energy. The energy required to activate the ion pumps is produced mainly from glucose carried by the blood. More blood flow is necessary to transport more glucose, also bringing in more oxygen in the form of oxygenated hemoglobin molecules in red blood cells. The blood-flow increase happens within 2–3 millimeters of the active neurons. Usually the amount of oxygen brought in exceeds the amount of oxygen consumed in burning glucose, which causes a net decrease in deoxygenated hemoglobin in that area of a brain. Although one might expect blood oxygenation to decrease with activation, the dynamics are a bit more complex than that. There is indeed a momentary decrease in blood oxygenation immediately after neural activity increases, but it is followed by a period during which the blood flow increases, overcompensating for the increased demand, and blood oxygenation actually increases after neuronal activation. That phenomenon, first reported by Seiji Ogawa, is known as the *blood-oxygenation-level-dependent (BOLD) effect*. It changes the properties of the blood near the firing neurons. Because it provides information about the level of blood flow in different regions of the brain, it can be used to detect and monitor brain activity (Ogawa et al. 1990). The magnitude of the BOLD signal peaks after a few seconds and then falls back to a base level. There is evidence that the BOLD signal is more closely related to the input than to the output activity of the neurons in the region (Raichle and Mintun 2006). In parts of the cortex where the axons are short and their ends are near the neuron bodies, it makes no difference whether the BOLD signal is correlated with the input or with the output of the neurons, since the voxels are not small enough to distinguish between the different parts of the neuron. In other areas of the brain, where the axons are longer, the difference between input and output activity can be significant.

Obtaining accurate measures of the level of the BOLD signal is difficult, since the signal is weak and can be corrupted by noise from many sources. In practice, sophisticated statistical procedures are required to recover the underlying signal. The resulting information about brain activation can be viewed graphically by color coding the levels of activity in the whole brain or in the specific region being studied. By monitoring the BOLD signal, it is possible to localize activity to within millimeters with a time resolution of a few seconds. Alternative technologies that can improve both spatial resolution and time resolution through the use of biomarkers other than the BOLD signal are under development, but they have other limitations. Therefore, the majority of techniques used today use the BOLD signal as a proxy for brain activity.

Near-infrared spectroscopy (NIRS) is a technique based on the use of standard electromagnetic radiation that uses a different part of the spectrum than CT techniques use: the range from 100 to 400 terahertz (that is, from 1×10^{14} to 4×10^{14} Hz). NIRS can be used to study the brain because transmission and absorption of NIR photons by body tissues reveal information about changes in hemoglobin concentration (Villringer et al. 1993). NIRS can be used non-invasively to monitor brain function by measuring the BOLD signal because in the NIRS frequency range light may diffuse several centimeters through the tissue before it is diffused and detected (Boas, Dale, and Franceschini 2004). A NIRS measurement consists in sending photons of appropriate frequency into the human brain, sensing the diffused light, and using computer algorithms to compute the densities of substances causing the photons to diffuse.

NIRS is sensitive to the volume of tissue residing between the source of light entering the tissue and the detector receiving the light that diffuses out of the tissue. Since NIR light penetrates only a few centimeters into the human brain before being diffused, the source and the detector are typically placed on the scalp, separated by a few centimeters. The resulting signal can be used to image mainly the most superficial cortex. NIRS is a non-invasive technique that can be used to measure hemodynamic signals with a temporal resolution of 100 Hz or better, although it is always limited by the slow response of the BOLD effect. Functional NIR imaging (fNIR) has several advantages in cost and portability over MRI and other techniques, but it can't be used to measure cortical activity more than a few centimeters deep in the skull, and it has poorer spatial resolution. The use of NIRS in functional mapping of the human cortex is also called *diffuse optical tomography* (DOT).

Positron-emission tomography (PET) is a computerized imaging technique that uses the particles emitted by unstable isotopes that have been injected into the blood. The technique is based on work done by David Kuhl, Luke Chapman, and Roy Edwards in the 1950s at the University of Pennsylvania. In 1953, Gordon Brownell, Charles Burnham, and their group at Massachusetts General Hospital demonstrated the first use of the technology for medical imaging (Brownell and Sweet 1953).

PET is based on the detection of pairs of high-energy photons emitted in the decay of a positron emitted by a radioactive nucleus that has been injected into the body as part of a biologically active molecule. A positron is a sub-atomic particle with the same mass as an electron, but with positive charge. It is, in fact, a piece of antimatter—an anti-electron. When emitted from a decaying radionuclide integrated in ordinary matter, a positron can travel only a very short distance before encountering an electron, an event that annihilates both particles and results in two high-energy photons (511 KeV, or kilo-electron-volts) leaving the site of annihilation and traveling in opposite directions, almost exactly 180 degrees from one another. Three-dimensional images of the concentration of the original radionuclide (the tracer) within the body are then constructed by means of automated computer analysis.

The biologically active molecule most often chosen for use in PET is fluorodeoxyglucose (FDG), an analogue of glucose. When it is used, the concentrations of tracer imaged are correlated with tissue metabolic activity, which again involves increased glucose uptake and the BOLD signal. The most common application of PET, detection of cancer tissues, works well with FDG because cancer tissues, owing to their differences in structure from non-cancerous tissues, produce visible signatures in PET images.

When FDG is used in imaging, the normal fluorine atom in each FDG molecule is replaced by an atom of the radioactive isotope fluorine-18. In the decay process, known as β+ decay, a proton is replaced by a neutron in the nucleus, and the nucleus emits a positron and an electron neutrino. (See figure 9.1.) This isotope, ^{18}F, has a half-life (meaning that half of the fluorine-18 atoms will have emitted one positron and decayed into stable oxygen-18 atoms) of 110 minutes.

One procedure used in neuroimaging involves injecting labeled FDG into the bloodstream and waiting about half an hour so that the FDG not used by neurons leaves the brain. The labeled FDG that remains in the brain tissues has been metabolically trapped within the tissue, and its

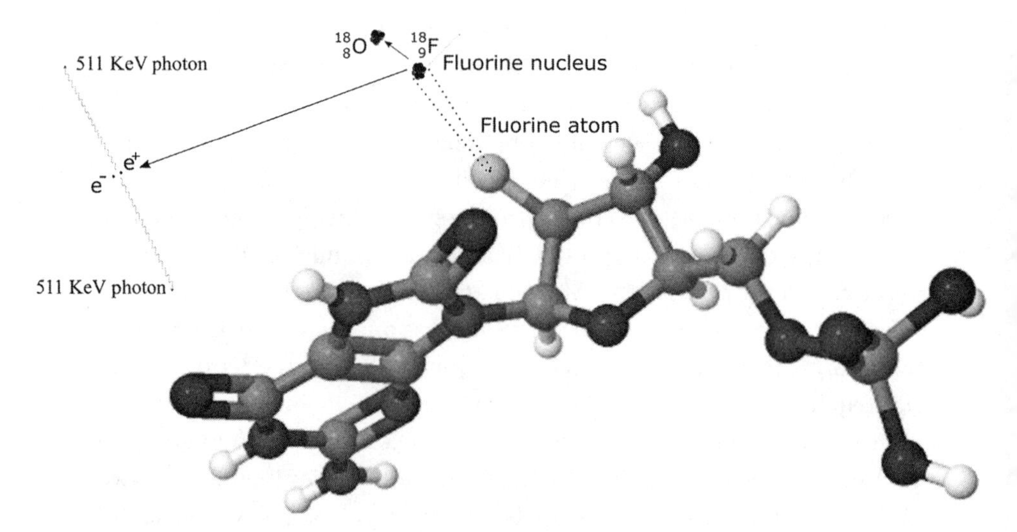

Figure 9.1
A schematic depiction of the decay process by which a fluorine-18 nucleus leads to the production of two high-energy photons (drawing not to scale).

concentration gives an indication of the regions of the brain that were active during that time. Therefore, the distribution of FDG in brain tissue, measured by the number of observed positron decays, can be a good indicator of the level of glucose metabolism, a proxy for neuronal activity. However, the time scales involved are not useful for studying brain activity related to changes in mental processes lasting only a few seconds.

However, the BOLD effect can be used in PET to obtain evidence of brain activity with higher temporal resolution. Other isotopes with half-lives shorter than that of FDG—among them carbon-11 (half-life 20 minutes), nitrogen-13 (half-life 10 minutes), and oxygen-15 (half-life 122 seconds) are used, integrated into a large number of different active molecules. The use of short-lived isotopes requires that a cyclotron be nearby to generate the unstable isotope so that it can be used before a significant fraction decays.

Despite the relevance of the previous techniques, the technique most commonly used today to study macro-scale brain behavior is magnetic-resonance imaging. Paul Lauterbur of the State University of New York at Stony Brook developed the theory behind MRI (Lauterbur 1973), building on previous work by Raymond Damadian and Herman Carr.

MRI is based on a physical phenomenon that happens when hydrogen nuclei are exposed to electric and magnetic fields. Hydrogen nuclei, which are in fact protons, have an intrinsic property, called *nuclear spin*, that makes them behave like small magnets that align themselves parallel to an applied magnetic field. When such a field is applied, a small fraction of the nuclei of atoms of hydrogen present, mostly in water molecules, align with the applied field. They converge to that alignment after going through a decaying oscillation of a certain frequency, called the *resonating frequency*, which depends on the intensity of the field. For a magnetic field of one tesla (T), the resonating frequency is 42 megahertz and it increases linearly with the strength of the field. Existing equipment used for human body imaging works with very strong magnetic fields (between 1.5 and 7 teslas). For comparison, the strength of the Earth's magnetic field at sea level ranges from 25 to 65 microteslas, a value smaller by a factor of about 100,000.

If one applies a radio wave of a frequency close enough to the resonating frequency of these nuclei, the alignment deviates from the applied magnetic field, much as a compass needle would deviate from the north-south line if small pushes of the right frequency were to be applied in rapid succession. When this excitation process terminates, the nuclei realign themselves with the magnetic field, again oscillating at their specific resonating frequency. While doing this, they emit radio waves that can be detected and processed by computer and then used to create a three-dimensional image in which different types of tissues can be distinguished by their structure and their water content.

In practice, it isn't possible to detect with precision the position of an atom oscillating at 42 megahertz, or some frequency of the same order, because the length of the radio waves emitted is too long. Because of physical limitations, we can only detect the source of a radio wave with an uncertainty on the order of its wavelength, which is, for the radio waves we are considering here, several meters. However, by modulating the magnetic field and changing its strength with time and along the different dimensions of space, the oscillating frequency of the atoms can be finely controlled to depend on their specific positions. This modulation causes each nucleus to emit at a specific frequency that varies with its location, and also with time, in effect revealing its whereabouts to the detectors. Figure 9.2 illustrates how control of the magnetic field can be used to pinpoint the locations of oscillating hydrogen nuclei.

MRI has some limitations in terms of its space and time resolution. Because a strong enough signal must be obtained, the voxels cannot be too small. The stronger the magnetic field, the smaller the voxels can be, but even 7-tesla MRI machines cannot obtain high-quality images with voxels much smaller than about one cubic millimeter. However, as technology evolves, one can hope that this resolution will improve, enabling MRI to obtain images with significantly smaller voxels.

The physical principles underlying MRI can be used in accordance with many different protocols to obtain different sorts of information. In its simplest and most commonly used form, MRI is used to obtain static images of the brain for the purpose of diagnosing tumors or other diseases, which manifest themselves as changes in the brain's macro-structures. These changes become visible because different tissues have different MRI signatures. Figure 9.3 shows the images of three slices of one brain. Image a corresponds to a slice near the top of the skull, with the cortex folds clearly visible. Image b shows some additional cortex folds; also visible are the lateral ventricles, where cerebrospinal fluid is produced, and the corpus

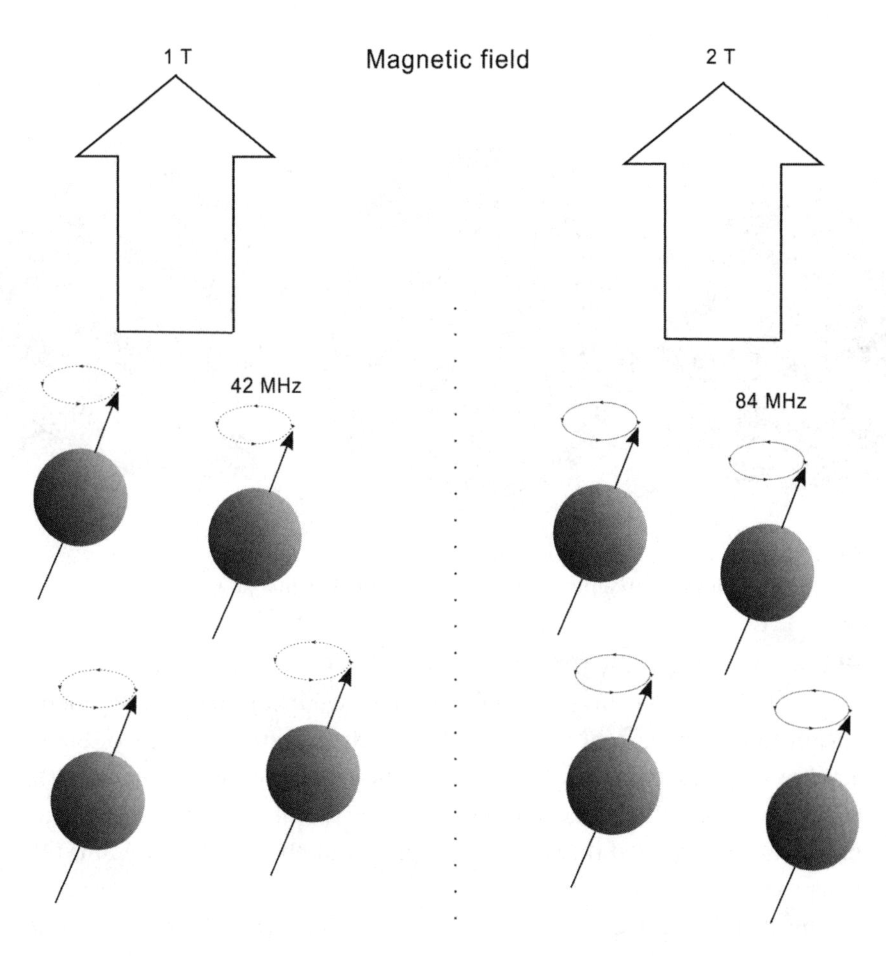

Figure 9.2
An illustration of how protons, oscillating at frequencies that depend on the strength of the magnetic field, make it possible to determine the precise locations of hydrogen atoms.

callosum, the largest white matter structure in the brain. Image c shows a clearly abnormal structure (a benign brain tumor) on the right side of the image, near the center.

MRI techniques can also be used to obtain additional information about brain behavior and structure. Two important techniques are functional MRI and diffusion MRI.

In recent decades, functional MRI (fMRI) has been extensively used in brain research. Functional MRI was first proposed in 1991 by Jack Belliveau,

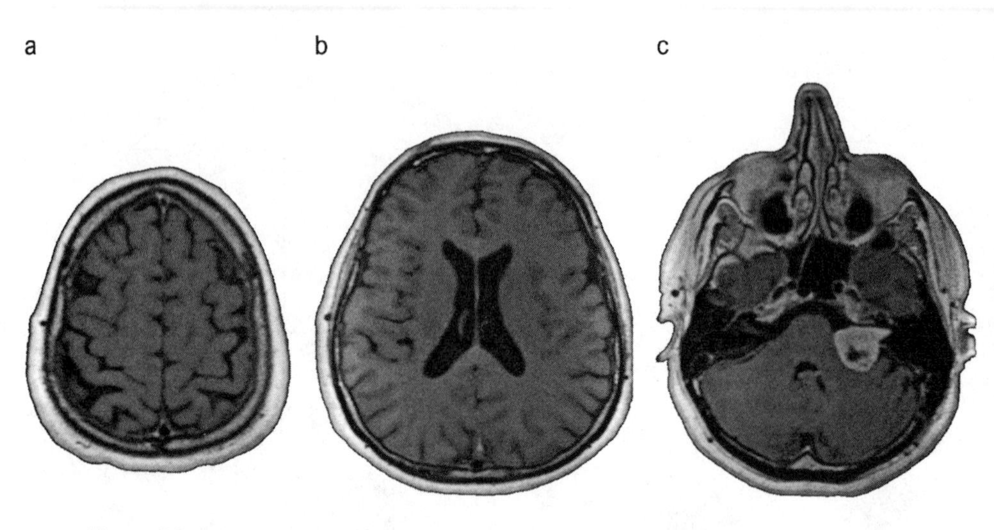

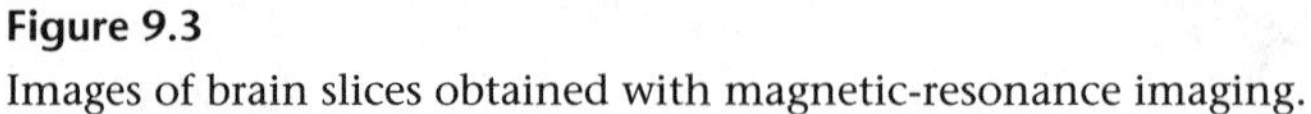

Figure 9.3
Images of brain slices obtained with magnetic-resonance imaging.

who was working at the Athinoula A. Martinos Center, in Boston. (See Belliveau et al. 1991.) Belliveau used a contrast agent injected in the bloodstream to obtain a signal that could be correlated with the levels of brain activity in particular areas. In 1992, Kenneth Kwong (Martinos Center), Seiji Ogawa (AT&T Bell Laboratories), and Peter Bandettini (Medical College of Wisconsin) reported that the BOLD signal could be used directly in fMRI as a proxy for brain activity.

The BOLD effect can be used in fMRI because the difference in the magnetic properties of oxygen-rich and oxygen-poor hemoglobin leads to differences in the magnetic-resonance signals of oxygenated and deoxygenated blood. The magnetic resonance is stronger where blood is more highly oxygenated and weaker where it is not. The time scales involved in the BOLD effect largely define the temporal resolution. For fMRI, the hemodynamic response lasts more than 10 seconds, rising rapidly, peaking at 4 to 6 seconds, and then falling exponentially fast. The signal detected by fMRI lags the neuronal events triggering it by a second or two, because it takes that long for the vascular system to respond to the neuron's need for glucose.

The time resolution of fMRI is sufficient for studying a number of brain processes. Neuronal activities take anywhere from 100 milliseconds to a few seconds. Higher reasoning activities, such as reading or talking, may take anywhere from a few seconds to many minutes. With a time resolution of a few seconds, most fMRI experiments study brain processes lasting from a few seconds to minutes, but because of the need for repetition, required to improve the signal-to-noise ratio, the experiments may last anywhere from a significant fraction of an hour to several hours,.

Diffusion MRI (dMRI) is another method that produces magnetic-resonance images of the structure of biological tissues. In this case, the images represent the local characteristics of molecular diffusion, generally of molecules of water (Hagmann et al. 2006). Diffusion MRI was first proposed, by Denis Le Bihan, in 1985. The technique is based on the fact that MRI can be made sensitive to the motion of molecules, so that it can be used to show contrast related to the structure of the tissues at microscopic level (Le Bihan and Breton 1985). Two specific techniques used in dMRI are diffusion weighted imaging (DWI) and diffusion tensor imaging (DTI).

Diffusion weighted imaging obtains images whose intensity correlates with the random Brownian motion of water molecules within a voxel of tissue. Although the relationship between tissue anatomy and diffusion is complex, denser cellular tissues tend to exhibit lower diffusion coefficients, and thus DWI can be used to detect certain types of tumors and other tissue malformations.

Tissue organization at the cellular level also affects molecule diffusion. The structural organization of the white matter of the brain (composed mainly of glial cells and of myelinated axons, which transmit signals from one region of the brain to another) can be studied in some detail by means of diffusion MRI because diffusion of water molecules takes places preferentially in some directions. Bundles of axons make the water diffuse preferentially in a direction parallel to the direction of the fibers.

Diffusion tensor imaging uses the restricted directions of diffusion of water in neural tissue to produce images of neural tracts. In DTI, to each voxel corresponds one ellipsoid, whose dimensions give the intensity and the directions of the diffusion in the voxel. The ellipsoid can be characterized by a matrix (a tensor). The directional information at each voxel can be used to identify neural tracts. By a process called *tractography,* DTI data can be used to represent neural tracts graphically. By color coding individual tracts, it is possible to obtain beautiful and highly informative images of the most important connections between brain regions.

Tractography is used to obtain large-scale, low-resolution information about the connection patterns of the human brain. Figure 9.4 shows an image obtained by means of diffusion tensor imaging of the mid-sagittal plane, the plane that divides the brain into a left and a right side. The axon fibers connecting different regions of the brain, including the fibers in the corpus callosum connecting the two hemispheres and crossing the mid-sagittal plane, are clearly visible.

Electroencephalography (EEG) and magnetoencephalography (MEG) are two techniques that obtain information about brain behavior from physical principles different from the methods discussed above. Both EEG and MEG measure the activity level of neurons directly by looking how that activity affects electromagnetic fields.

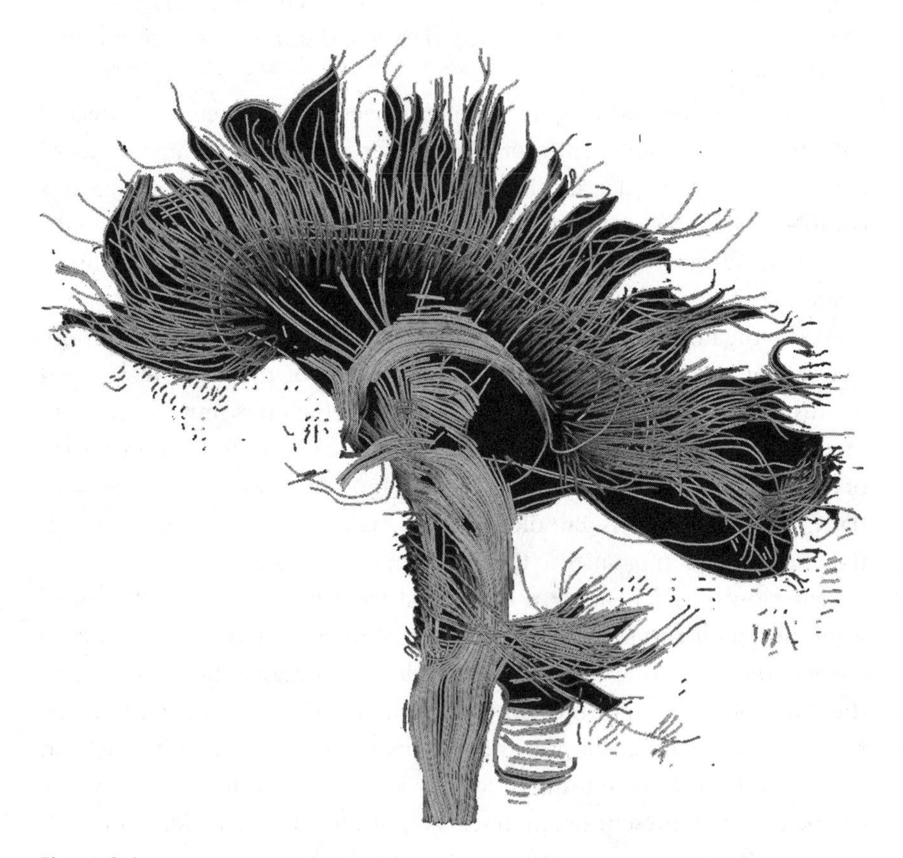

Figure 9.4
DTI reconstruction of tracts of brain fiber that run through the mid-sagittal plane. Image by Thomas Schultz (2006), available at Wikimedia Commons.

When large groups of neurons fire in a coordinated way, they generate electric and magnetic fields that can be detected. The phenomenon, discovered in animals, was first reported by Richard Caton, a Liverpool physician (Caton 1875). In 1924, Hans Berger, inventor of the technique now known as EEG, recorded the first human electroencephalogram. (See Berger 1929.)

EEG works by recording the brain's electrical activity over some period of time from multiple electrodes, generally placed on the scalp. The electric potential generated by the activity of an individual neuron is far too small to be picked up by the electrodes. EEG activity, therefore, reflects the contribution of the correlated activity of many neurons with related behavior and similar spatial orientation. Pyramidal neurons in the cortex produce the largest part of a scalp-recorded EEG signal because they are well aligned and because they fire in a highly correlated way. Because electric fields fall off with the square of the distance, activity from deeper structures in the brain is more difficult to detect.

The most common analyses performed on EEG signals are related to clinical detection of dysfunctional behaviors of the brain, which become visible in EEG signals as abnormal patterns. However, EEG is also used in brain research. Its spatial resolution is very poor, since the origin of the signals can't be located more precisely than within a few centimeters. However, its time resolution is very good (on the order of milliseconds). EEG is sometimes used in combination with another imaging technique, such as MRI or PET.

In a related procedure known as electrocorticography (ECoG) or intracranial EEG (iEEG), electrodes are placed directly on the exposed surface of the brain to record electrical activity from the cerebral cortex. Since it requires removing a part of the skull to expose the brain's surface, ECoG is not widely used in fundamental brain research with healthy human subjects, although it is extensively used in research with animals.

Magnetoencephalography (MEG), first proposed by David Cohen (1968), also works by detecting the effects in the electromagnetic fields of firing neurons, but in MEG the detectors are very sensitive magnetometers. To reduce the magnetic background noise, Cohen used a magnetically shielded room. (See figure 9.5.)

Whereas EEG detects changes in the electric field caused by ion flows, MEG detects tiny changes in the magnetic field caused by the

Figure 9.5
A model of the magnetically shielded room built by David Cohen in 1969 at MIT for the first magnetoencephalography experiments. Photo taken at Athinoula A. Martinos Center for Biomedical Imaging.

intra-neuron and extra-neuron currents that occur in the correlated firing of large numbers of neurons (Okada 1983). Since currents must have similar orientations to generate magnetic fields that reinforce one another, it is again (as in EEG) the layer of pyramidal cells, which are situated perpendicular to the cortical surface, that produce the most easily measurable signals. Bundles of neurons with an orientation tangential to the scalp's surface project significant portions of their magnetic fields outside the head. Because these bundles are typically located in the sulci, MEG is more useful than EEG for measuring neural activity in those regions, whereas EEG is more sensitive to neural activity generated on top of the cortical sulci, near the skull.

MEG and EEG both measure the perturbations in the electromagnetic field caused by currents inside the neurons and across neuron cell walls, but they differ greatly in the technology they use to do so. Whereas EEG uses relatively cheap electrodes placed on the skull, each connected to one amplifier, MEG uses arrays of expensive, highly sensitive magnetic detectors called *superconducting quantum interference devices* (SQUIDS). An MEG detector costs several million dollars to install and is very expensive to operate. Not only are the sensors very expensive; in addition, they must be heavily shielded from external magnetic fields, and they must be cooled to ultra-low temperatures. Efforts are under way to develop sensitive magnetic detectors that don't have to be cooled to such low temperatures. New technologies may someday make it possible to place magnetic detectors closer to the skull and to improve the spatial resolution of MEG.

A recently developed method known as *event-related optical signal* (EROS) uses the fact that changes in brain-tissue activity lead to different light-scattering properties. These changes may be due to volumetric changes associated with movement of ions and water inside the neurons and to changes in ion concentration that change the diffraction index of the water in those neurons (Gratton and Fabiani 2001). Unlike the BOLD effect, these changes in the way light is scattered take place while the neurons are active and are spatially well localized. The spatial resolution of EROS is only slightly inferior to that of MRI, but its temporal resolution is much higher (on the order of 100 milliseconds). However, at present EROS is applicable only to regions of the cortex no more than a few centimeters away from the brain's surface.

The techniques described in this section have been extensively used to study the behavior of working brains, and to derive maps of activity that provide significant information about brain processes, at the macro level. Whole-brain MRI analysis has enabled researchers to classify the voxels in a working brain into several categories in accordance with the role played by the neurons in each voxel (Fischl et al. 2002; Heckemann et al. 2006). Most studies work with voxels on the order of a cubic millimeter, which is also near the resolution level of existing MRI equipment. Many efforts aimed at making it possible to integrate imaging data from different subjects, and a number of atlases of the human brain have been developed—for example, the Probabilistic Atlas and Reference System for the Human Brain (Mazziotta et al. 2001) and the BigBrain Atlas (developed in the context of the Human Brain Project).

Integration of different techniques, including MRI, EEG, PET, and EROS, may lead to an improvement of the quality of the information retrieved. Such integration is currently a topic of active research.

A recently developed technique that shows a lot of promise to advance our understanding of brain function is *optogenetics*. In 1979, Francis Crick suggested that major advances in brain sciences would require a method for controlling the behavior of some individual brain cells while leaving the behavior of other brain cells unchanged. Although it is possible to electrically stimulate individual neurons using very thin probes, such a method does not scale to the study of large numbers of neurons. Neither drugs nor electromagnetic signals generated outside the brain are selective enough to target individual neurons, but optogenetics promises to be. Optogenetics uses proteins that behave as light-controlled membrane channels to control the activity of neurons. These proteins were known to exist in unicellular algae, but early in the twenty-first century researchers reported that they could behave as neuron membrane channels (Zemelman et al. 2002; Nagel et al. 2003) and could be used to stimulate or repress the activity of individual neurons or of smalls groups of neurons, both in culture (Boyden et al. 2005) and in live animals (Nagel et al. 2005). By using light to stimulate specific groups of neurons, the effects of the activity of these neurons on the behavior of the organism can be studied in great detail.

Optogenetics uses the technology of genetic engineering to insert into brain cells the genes that correspond to the light-sensitive proteins. Once these proteins are present in the cells of modified model organisms, light of the appropriate frequency can be used to stimulate neurons in a particular region. Furthermore, the activity of these neurons can be controlled very precisely, within milliseconds, by switching on and off the controlling light. Live animals, such as mice or flies, can be instrumented in this way, and the behaviors of their brains can then be controlled in such a way as to further our understanding of brain circuitry.

Optogenetics is not, strictly speaking, an imaging technique, but it can be used to obtain detailed information about brain behavior in live organisms with a resolution that cannot be matched by any existing imaging technique.

Brain Networks

The imaging technologies described in the previous section, together with other information, enabled researchers to study how various regions of the brain are connected, both structurally and functionally. They provide extensive information about the axon bundles that connect regions of the brain and about the levels of activity in various regions when the brain is performing specific tasks.

In studying brain networks, each small region of the brain, defined by a combination of geometrical, physiological, and functional considerations, is made to correspond to one node in a graph. Subjects are then asked either to perform a specific task or to stay in their resting state while the brain is imaged and the level of activity of each of these regions is recorded. Statistical analyses of the data and graph algorithms are then used to study different aspects of the brain's behavior and structure. When more than one subject is used in a study, the coordinates of the different brains are mapped into a common coordinate system; this is necessary because individual brains differ in their geometry and hence simple, fixed spatial coordinates don't refer to the same point in two different brains. Linear and non-linear transformations are therefore used to map the coordinates of different brains into the coordinates of one reference brain, which makes it possible to obtain a fairly close correspondence between the equivalent brain regions in different individuals.

As an illustration, let us consider one particular brain network obtained from high-resolution fMRI data. In an experiment conducted collaboratively by researchers from the Martinos Center and Técnico Lisboa, one of the objectives was to identify the strongest functional connections between brain regions. The brains of nine healthy subjects, who were at rest, were imaged on a high-field (7 teslas) whole-body scanner with an isotropic spatial resolution of 1.1 millimeter and a temporal resolution of 2.5 seconds. For different subjects, a number of regions of interest (ROI) were selected, using a method whose purpose was to identify stable and significant ROIs (Dias et al. 2015). The correlations between the levels of activity in these regions were then used to construct a graph, with the edge weights given by the value of the correlations. The graph was processed by identifying the maximum spanning tree, then adding to the tree the strongest remaining connections between ROIs, in decreasing order of their strength, until the average node degree reached 4 (Hagmann et al. 2008). The maximum spanning tree can be found by adapting the method that was used to find the minimum spanning tree, using the algorithm discussed in chapter 4. This procedure yields a network showing 124 regions of interest (figure 9.6). Higher numbers of ROIs lead to graphs that are more informative but are more difficult to interpret visually.

The same fMRI data were also used to study the structure and characteristics of very large graphs derived by considering one ROI for each voxel imaged in each subject. Such graphs have more than a million nodes and up to a billion edges, depending on the minimum level of correlation used to define a connection between ROIs (Leitão et al. 2015).

Hundreds of methodologies have been used to study brain networks in many thousands of studies in diverse fields including psychology, neurology, genetics, and pharmacology. Brain networks can be used to study neural circuits and systems, the firing patterns that characterize specific activities or sensations, brain diseases, and many other features of brain behavior. Olaf Sporns' book *Networks of the Brain* covers many of the techniques and objectives of brain network studies and is an excellent reference work on the subject.

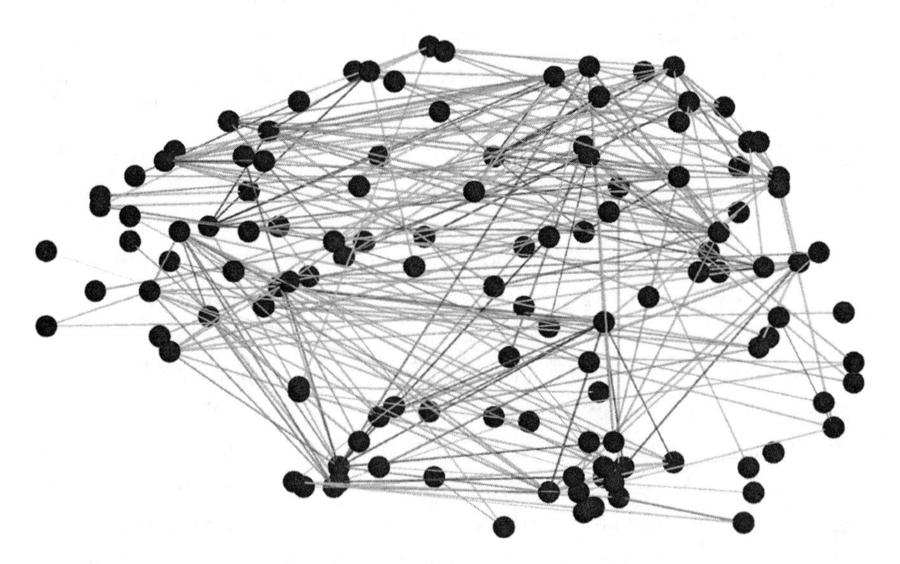

Figure 9.6
A brain network obtained by considering the highest correlations among 124 different regions of interest in an average of nine human subjects.

Slicing and Dicing

Since the imaging methods described above have limited spatial resolution, fine-grained information about neuron structure and connectivity can more easily be obtained from dead brains, which can be sliced and observed at the microscope in great detail. Once slices of the brain have been obtained, microscopy doesn't destroy the samples and can provide a spatial resolution sufficient to resolve individual neurons, axons, dendrites, synapses, and other structures.

In conventional microscopy, which is derived directly from the technology used in sixteenth-century microscopes, the entire sample is flooded with light, and thus all parts of the sample in the optical path are excited at the same time. Reflection microscopy and transmission microscopy measure the reflected or transmitted light directly, with a resolution limited by a number of factors (most notably the wavelength of the light used to image the samples). Light visible to the human eye has wavelengths between 400 and 700 nanometers, which means that features much smaller than one micrometer (μm) cannot be resolved. Neuron cells' bodies range in size from 4 to 50 μm, but synapses are much smaller, filling a gap between neurons that is only 20–40 nm wide (Hubel 1988). Therefore, resolving individual neuron components, such as synapses requires electron microscopy.

Electron microscopy uses electrons instead of photons to image samples. Because the wavelength of an electron can be shorter than the wavelength of a visible-light photon by a factor of 10^3, an electron microscope has a much higher resolution than an optical microscope. Electron microscopy can resolve features smaller than 0.1 nm, several orders of magnitude smaller than the resolution of optical microscopes. Electron microscopy has, therefore, been used extensively to determine the detailed structures of individual neurons and of small groups of neurons. Electron microscopy, however, suffers from several limitations on its use in biological tissues (it is especially difficult to use in tissues that are kept alive). Other methods can be used to observe living tissues, or tissues stained with biological markers related to specific biological processes.

In fluorescent microscopy, what is imaged is not reflected light but light emitted by a sample that has absorbed light or some other form of electromagnetic radiation. The sample must include in its composition fluorophores—fluorescent chemical compounds that re-emit light upon being excited by visible radiation. The wavelength of the fluorescent light is different from the wavelength of the absorbed radiation, and depends on the nature of the fluorophores. The resulting fluorescence is detected by the microscope's photo-detector. Since cells can be labeled with different fluorescent molecules related to different biological processes, fluorescent microscopy can be used to trace specific proteins inside cells.

Confocal microscopy (Minsky 1961) uses point illumination and a pinhole close to the sample to eliminate out-of-focus light coming from other parts of the sample. Because only light produced by fluorescence very close to the focal plane can be detected, the image's optical resolution, particularly in the depth direction, is much better than that of wide-field microscopes.

Multi-photon fluorescent microscopy (MFM) is another technique that has extended the range of application of optical microscopy. Like fluorescent microscopy, MFM uses pulsed long-wavelength light to excite fluorophores within a specimen. In MFM, the fluorophore must absorb the energy of several long-wavelength photons, which must arrive almost at the same time, in order to excite an electron into a higher energy state. When the electron comes back to its ground state, it emits a photon. By controlling the laser sources, it is possible to image live cells in better conditions than the conditions that are possible when using alternative techniques. Two-photon microscopy is a special case of multi-photon microscopy in which exactly two photons of infrared light are absorbed by the fluorophore.

Whatever the microscopy technique used, only very thin slices of brain tissue can be imaged, because it isn't possible to image sections deep inside the slice. The slices imaged can correspond to slices cut from the surface of the tissue, or to the top layer of a block of tissue.

Slices of brain tissue that have been chemically hardened to make them amenable to slicing can be cut with a microtome or an ultramicrotome. Basically glorified meat slicers, such devices can cut very thin slices of samples. Steel, glass, or diamond blades can be used, depending on the material being sliced and the desired thickness of the slices. Steel and glass blades can be used to prepare sections of animal or plant tissues for light microscopy. Diamond blades are used to prepare thin sections of brain for high-resolution microscopy.

Microtomes have been used for decades to obtain two-dimensional images of brain slices. Researchers seeking to obtain more detailed three-dimensional information about neuron arrangement have taken to imaging stacks of slices and combining the information to obtain complete three-dimensional information. That technique, called serial electron microscopy (SEM), was used to obtain the whole structure of the worm *C. elegans* in a laborious process that took many years. The difficulty of using SEM lies in the fact that, as slices are removed and put into glass blades, they become distorted and difficult to align. Despite these difficulties, SEM has been extensively used to obtain information about three-dimensional brain structures.

A significant improvement over serial electron microscopy came when Winfried Denk and Heinz Horstmann (2004) proposed serial block-face electron microscopy (SBEM). In SBEM, the face of a whole block of tissue is imaged by putting the block inside an electron microscope. Then a thin slice of the block is removed with a diamond blade microtome, and is discarded. The newly exposed slice is then imaged again. The process can be automated and obtains images with less distortion than does SEM.

Images obtained by SEM and by SBEM can be processed to reveal the three-dimensional geometry of individual neurons and even the locations of individual synapses. This is a process essentially similar to computed tomography. Together, two-dimensional imaging and computerized combination of images have led to many different techniques now used in various circumstances and settings. Many different combinations of microscopy techniques, tissue preparation, serial slice analysis, and computerized image processing have been used by researchers to obtain very detailed information about larger and larger brain structures.

Conceptually, the whole structure of connections in a whole brain can be mapped, even though the state of the art still imposes significant limitations on what can be done. A number of efforts to map large regions of mammalian brains have been undertaken by researchers, and the data they have obtained have been made available in the Open Connectome Project.

In 2011, researchers from Harvard University and Carnegie Mellon University used two-photon calcium imaging (a technique used to monitor the activity of neurons in live brain tissue) and serial electron microscopy to obtain both functional and structural information about a volume of the mouse visual cortex measuring 450 × 350 × 52 μm and containing approximately 1,500 neurons (Bock et al. 2011). The researchers used two-photon calcium imaging to locate thirteen neurons that had responded to a particular stimulus, then reconstructed a graph of connections of those thirteen neurons.

In 2013, a group of researchers reported the reconstruction of 950 neurons and their mutual contacts in one section of the retina of a mouse (Helmstaedter et al. 2013). They used serial block-face electron microscopy to acquire sections from an 114-by-80-μm area of the mouse retina. The data were annotated by human curators and by machine learning algorithms to yield the full structure of that region of the mouse retina. Helmstaedter et al. estimated that more than 20,000 annotator hours were spent on the whole process.

In work reported in 2015,, researchers from Harvard, MIT, Duke University, and Johns Hopkins University fully reconstructed all the neuron sections and many sub-cellular objects (including synapses and synapse vesicles) in 1,500 cubic micrometers (just a little more than a millionth of a cubic millimeter) of mouse neocortex, using 3 × 3 × 30 nm voxels (Kasthuri et al. 2015). They obtained 2,250 brain slices, each roughly 30 nm thick, using a tape-collecting ultramicrotome equipped with a diamond blade. The slices were imaged by serial electron microscopy; then the images were processed in order to reconstruct a number of volumes. In these volumes, the authors reconstructed the three-dimensional structure of roughly 1,500 μm^3 of neural tissue, which included hundreds of dendrites, more than 1,400 neuron axons, and 1,700 synapses (an average of about one synapse per cubic micron). Figure 9.7, reprinted from one of the articles that describe this effort, shows a rendering of one small cylinder of neocortex.

The efforts described above and many other efforts have shown that it is possible to obtain highly detailed information about fine-grained brain structure. Such information may, in time, be sufficiently detailed and reliable to define the parameters required to perform accurate neuron-level simulation of brain sections.

Brain Simulation

As our understanding of the mechanisms and structures of living brains evolves, it may become possible to use engineering approaches to replicate

Figure 9.7
A rendering of one cylindrical volume of neocortex, roughly 8 μm in diameter and 20 μm long, obtained from the reconstruction performed by Kasthuri et al. (2015). Reprinted with permission from Narayanan Kasthuri.

and to study in great detail several different phenomena in the brain. When the fine structure of brain tissue and the mechanisms that control brain development are understood in detail, it will become possible to simulate brain development and behavior. The simulations will have to reproduce the complex biochemical mechanisms that direct neuron growth, the chemical and electrical basis for synapse plasticity (and other forms of brain plasticity), and, eventually, the detailed workings of the billions of neurons and trillions of synapses that constitute a working brain. However, if good models exist for these components, there is no fundamental reason why simulation of large regions of a working brain cannot be performed.

It is interesting to assess what the simulation of a large part of the brain, or even the emulation of a whole brain, would imply in computational terms. One of the most ambitious such effort to have been completed so far probably is the Blue Brain Project, a collaboration between the École Polytechnique Fédérale de Lausanne and IBM. The project started in 2006, and the initial goal was to simulate in a supercomputer one neocortical column of a rat, which can be viewed as the smallest functional unit of the mammalian neocortex. A neocortical column is about 2 mm tall, has a diameter of 0.5 mm, and contains about 60,000 neurons in a human and 10,000 neurons in a rat. A human cortex may have as many as 2 million such columns. The computing power needed to perform that task was considerable, as each simulated neuron requires computing power roughly equivalent to that of a laptop computer. However, supercomputing technology is rapidly approaching a point at which simulating large parts of the brain will become possible.

In the most recent work, the results of which were published in 2013, the model and the simulations used the software and hardware infrastructure of the Blue Brain Facility to model a cortical column of a rat, with 12,000 neurons and 5 million dendritic and somatic compartments. The simulations were performed using publicly available software, the NEURON package, running on a Blue Gene P supercomputer with 1,024 nodes and 4,096 CPUs (Reimann et al. 2013). Since so far we have no detailed structural data on the cortical columns of a rat, statistical information about neuron connectivity and synapse distribution was used to define the parameters of the model. We can extrapolate the results of that simulation to calculate the amount of computer power required to simulate more complex systems and even a complete brain. Of course, there is no conclusive evidence that the level of detail used to simulate this "small" subsystem of a rat's brain is adequate for more complex and powerful simulations, even though Reimann et al. argued that a behavior consistent with reality was observed in this particular simulation. However, let us assume, for the moment, that this level of detail is indeed appropriate and could be used to simulate accurately more complex systems.

The Blue Gene P supercomputer, in the configuration used in the aforementioned study, has been rated as able to perform about 14 teraFLOPS (trillion floating-point operations per second). That speed enabled the computer to simulate about 4 seconds of cortical activity in 3 hours of computer time, in effect simulating the system approximately 2,700 times slower than real time. The fastest computer available at the time of this writing—the Sunway TaihuLight supercomputer, with 10 million processor cores—has been rated at 93,000 teraFLOPS (93 petaFLOPS), about 6,600 times the speed of the Blue Gene P supercomputer used in the simulation mentioned above.

If we ignore problems related to the difficulty of scaling up the simulation to a computer with a much larger number of cores (problems that are real but can be addressed), then, in principle, the Sunway TaihuLight supercomputer could simulate, more than two times faster than real time, a 12,000-neuron cortical column of a rat. With all other things assumed equal (even though we know they probably are not), if we had a complete model of a human brain such a machine could simulate its 86 billion cells at a speed roughly 3 million times slower than real time. To simulate 10 seconds of real brain time (enough to simulate, say, the utterance of one small sentence) would require about one year of computer simulation. Even more striking is the difference in the amounts of energy needed by the supercomputer and by an actual brain. A brain, which uses about 20 watts of power (Drubach 2000), would spend about 200 joules to work for 10 seconds, whereas the 15MW Sunway TaihuLight would spend roughly 5×10^{14} joules to simulate the same event during this year, consuming 2 trillion times as much energy as the brain would.

The technologies involved, however, are evolving rapidly, making the simulations more and more efficient. The most ambitious simulation of a network of spiking neurons performed to date, developed to test the limits of the technology (Kunkel et al., 2014), used a supercomputer to emulate a network of 1.8 billion neurons, with an average number of 6000 synapses per neuron. The model used is not as detailed as the model used in the Blue Brain project, leading to a faster simulation at the expense of some precision.

The simulation used 40 minutes of the 8 petaFLOPS K computer, the fourth-fastest computer in the world at the time, to model the behavior of one second or real time of this large network of spiking neurons. Extrapolating to the size of a whole brain, and assuming no negative impact from the scaling up, the simulation of all the neurons in a human brain would still proceed 100,000 times slower than real time. The Sunway TaihuLight would have performed such a simulation "only" 10,000 times slower than real time, if all other parameters remained comparable.

The first reaction to the numbers cited above may be a combination of surprise, despair, and relief based on the renewed understanding—and now on comparatively hard data—that the human brain is simply too complex, too powerful, and too efficient to be fully simulated in a computer. In fact, even if we could obtain a working model of a brain (far in the future), and even if the fastest computer on Earth, allocated only to that task, could simulate such a model, the simulation would proceed many times slower than real time. This simple reasoning seems to show that such an endeavor will be forever impossible. However, because of two factors that cannot be ignored, that apparently obvious and clear conclusion is too naive.

The first of those two factors is the exponential growth in the power of computers, which hasn't stopped and which is likely to continue for quite a while. If Moore's Law continues to define the evolution of computers, in practice doubling their performance every two years, and if that rate of evolution is also true for the largest supercomputers, then a computer with the power to perform real-time simulation of a human brain will be available around the year 2055. There are a number of assumptions in these computations that may well prove to be false, including the sustainability of the evolution predicted by Moore's Law. Nonetheless, it remains likely that such a computer will exist in 40 or 50 years.

The second factor is, in my opinion, even more important than the first. It was noted in chapter 7 that simulation of digital chips is performed using multi-level simulation in which details of the behaviors of devices are abstracted away at the higher levels of the simulation. This means that systems with billions of transistors can be effectively simulated with multi-level models of the detailed workings of each individual transistor. The utilization of multi-level simulation techniques increases the speed of simulation by many orders of magnitude. There is no essential reason why such an approach cannot be effectively applied to the simulation of a complete brain. In fact, a number of authors—among them Marasco, Limongiello, and Migliore (2013)—have proposed ways to accelerate the simulation of neurons that would result in significant speedups. It is true it is much easier to create a multi-level model of an integrated circuit, with its well-defined modules and well-understood structure, than to create a multi-level model of a brain, with its complex network of connections and poorly defined modules. I believe, however, that the majority of the advances in this area are still to come. As more and more effort is put into modeling the behaviors of parts of the brain, more and more resources will be needed to run simulations. Multi-level abstractions of more complex brain blocks will be created and will be used to create higher-level models of brain regions—models that can be simulated much more rapidly in a computer without sacrificing any of the precision that is essential for the accurate simulation of a working brain.

Armed with a better understanding of computers, biological systems, and brains, let us now proceed to the central question that is the main topic of this book: Will we be able to create digital minds?

10 Brains, Minds, and Machines

The preceding chapters focused on the behavior of the one system known that undoubtedly creates a mind: the human brain. The question of whether the human brain is the only system that can host a mind is worth discussing in some detail.

Having a mind is somehow related to having an intelligence, although it may be possible to envision mindful behavior without intelligence behind it and intelligent behavior without a mind behind it. Let us assume, for the sake of discussion, that we accept the premises of the Turing Test, and that we will recognize a system as intelligent if it passes some sophisticated version of that test. We may defer to a later time important questions related to the level of sophistication of such a test, including how long the interaction would last, which questions could be asked, how knowledgeable the interrogators would have to be, and many additional details that are relevant but not essential to this discussion.

It is reasonable to think that a system that passes the Turing Test, and is therefore deemed intelligent, will necessarily have a mind of its own. This statement raises, among other questions, the question "What does it mean to have a mind?" After all, it is perfectly possible that a system behaves as an intelligent system, passes the Turing Test, yet doesn't have a mind of its own but only manages to fake having one. This question cannot be dismissed lightly. We will return to it later in this chapter.

In this book the word *mind* has already been used many times, most often in phrases such as "keep in mind" or in references to the "human mind." These are familiar uses of the word *mind*, and they probably haven't raised any eyebrows. However, anyone asked to define the meaning of the word *mind* more exactly will soon recognize that the concept is very slippery. The most commonly accepted definition of *mind* is "an emergent property of brain behavior that provides humans with a set of cognitive faculties, which include intelligence, consciousness, free will, reasoning, memory, and emotions." We may discuss whether other higher animals, such as dogs and monkeys, have minds of their own, and we may be willing to concede that they do but that their minds are different from and somewhat inferior to our own.

Perhaps the most defining characteristic of a mind is that it provides its owner with consciousness and with all that consciousness entails. After all, we may grant that a program running in a computer can reason, can perform elaborate mathematical computations, can play chess better than a human champion, can beat humans in TV game shows, can translate speech, and can even drive vehicles, but we would still be hard pressed to believe that a program with those abilities has free will, that it is self-conscious, that it has the concept of self, and that it fears its own death. We tend to view consciousness, in this sense, as a thing reserved for humans, even though we cannot pinpoint exactly what it is and where it comes from. Thousands of brilliant thinkers have tackled this problem, but the solution remains elusive.

Experience has made us familiar with only one very particular type of mind: the human mind. This makes it hard to imagine or to think about other types of minds. We may, of course, imagine the existence of a computer program that behaves and interacts with the outside world exactly as a human would, and we may be willing to concede that such a program may have a mind; however, such a program doesn't yet exist, and therefore the gap separating humans from machines remains wide.

Using a Turing Test to detect whether a computer is intelligent reveals an anthropocentric and restricted view of what a mind is. The Turing Test, which was conceived to remove anthropocentric predispositions, is still based on imitation of human behavior by a machine. It doesn't really test whether a program has a mind; it tests only whether a program has a mind that mimics the human mind closely.

Many thinkers, philosophers, and writers have addressed the question of whether a non-human mind can emerge from the workings of a program running in a computer, in a network of computers, or in some other computational support. But until we have a more solid definition of what a mind is, we will not be able to develop a test that will be able to recognize whether a given system has a mind of its own—especially if the system has a mind very different from the human minds we are familiar with.

To get around the aforementioned difficulty, let us consider a more restricted definition of a mind, which I will call a *field-specific* mind: a property of a system that enables the system to behave intelligently in a human-like way in some specific field. For instance, a program that plays chess very well must have a field-specific mind for chess, and a program that drives vehicles well enough to drive them on public streets must have a field-specific mind for driving. Field-specific minds are, of course, more limited than human minds. However, in view of the inherent complexity of a system that can behave intelligently and in a human-like way in any reasonably complex field, it seems reasonable to ask the reader to accept that such a system must have some kind of a field-specific mind. Such a field-specific mind probably is simpler than a human mind and probably doesn't give its owner consciousness, free will, self-awareness, or any of the characteristics that presumably emanate from self-awareness, such as fear of death.

A field-specific mind is an emergent property of a system that enables it to use some internal model of the world sufficiently well for the system to behave intelligently in some specific field and to act as intelligently and competently as a human in that field. It isn't difficult to imagine a field-specific intelligence test in which a system is tested against humans and is deemed to have a field-specific mind if it passes the test. The most stringent test of that kind would be a Turing Test in which a system not only has to behave intelligently in many specific fields but also must emulate human-like intelligence and behavior in all those fields.

Obviously, there are fields so specific and simple that having a field-specific mind for them is trivial. For instance, a system that never loses at tic-tac-toe will have a very simple field-specific mind, because tic-tac-toe is so easy that a very simple set of rules, encoded in a program (or even in a set of wires and switches), is sufficient to play the game. By the same token, a system that can play chess by following the rules, but that loses every game, has only a very simple mind for chess, close to no mind at all. On the other hand, a system that plays world-class chess and defeats human champions most of the time must have a field-specific mind for chess—there is no way it could do it by following some simple set of rules. With this proviso, we can accept that even creatures as simple as worms have field-specific minds that enable them to live and to compete in their particular environments, even if those minds are very simple and emanate from the workings of a very simple brain.

These definitions, somehow, define an ontology and a hierarchy of minds. A mind *A* can be viewed as more general than a mind *B* if it can be used to reason about any field in which mind *B* can also be used, and perhaps in other fields too. The human mind can be viewed as the most general mind known to date—a mind that encompasses many field-specific minds and that may exhibit some additional characteristics leading to self-awareness and consciousness. What the additional characteristics are and where they come from are important questions; they will be addressed later in this chapter.

I will now propose a classification of minds in accordance with their origins, their computational supports, and the forms of intelligence they exhibit. I will call a mind *synthetic* if it was designed and didn't appear "naturally" through evolution. I will call a mind *natural* if it has appeared through evolution, such as the human mind and, perhaps, the minds of other animals. I will call a mind *digital* if it emanates from the workings of a digital computer program. I will call a mind *biological* if it emanates from the workings of a biological brain. These considerations lead to the taxonomy illustrated in figure 10.1.

We are all familiar with natural minds, designed by evolution. Synthetic minds, designed by processes other than evolution to fulfill specific needs,

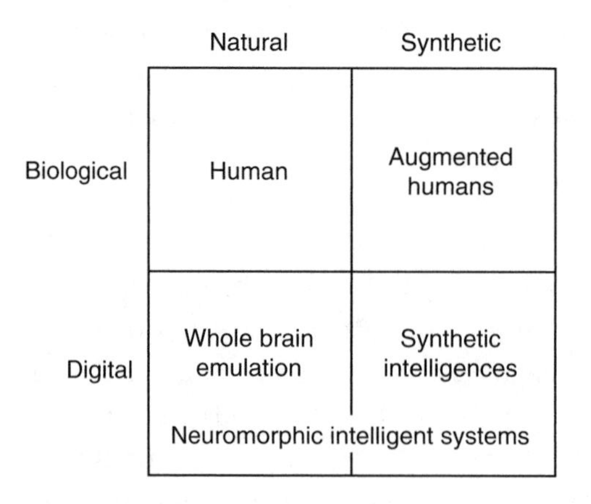

Figure 10.1
Digital and biological minds, natural and synthetic.

are likely to be developed in the next few decades. They probably will use digital supports, although if developed using the tools of synthetic biology (discussed in chapter 7) they could conceptually be supported by biological systems.

The entries in figure 10.1 may not be all equally intuitive. Almost everyone is familiar with natural, biological minds, the most obvious being the human mind. Many people are also familiar with the concept of synthetic, digital minds designed using principles that have nothing to do with how the brain operates. ("Synthetic intelligences" fall into the latter category.) I have also discussed the possibility that intelligent systems may someday be obtained by simulating the principles of brain development processes in a digital computer. I call such systems "neuromorphic intelligent systems." Additionally, rather than designing systems inspired by the principles at work in the brain, we may one day have the technology to simulate, or emulate, in minute detail, the behavior of a working human brain. That approach, called *whole-brain emulation*, corresponds to a natural mind (since it would work exactly as a human mind, which was designed by evolution) with a digital substrate.

It also is possible that we may one day be able to create synthetic minds with a biological substrate. The techniques of synthetic biology and advanced genetic engineering may someday be used to engineer beings with superhuman thinking abilities.

With this ontology of minds, we can now try to answer the central question of this book: What types of digital minds—field-specific or general, natural or synthetic—will, in the coming decades, emerge from the workings of intelligent programs running in digital computers?

Synthetic Intelligences

It is easy to identify present-day systems that were created to exhibit complex field-specific minds. By accessing large amounts of information, and by using sophisticated algorithms, these systems perform tasks that clearly require advanced field-specific minds. Such systems, developed by large teams of programmers, perform complex tasks that would require significant amounts of intelligence if they were to be performed by humans. I believe that they are the precursors of general-purpose synthetic minds, and that someday they will be recognized as such. As these systems evolve, they probably will become more and more intelligent until they become synthetic intelligences.

Deep Blue is a chess-playing computer developed by IBM. Its predecessor, Deep Thought, had been developed at Carnegie Mellon University. The creators of Deep Thought were hired by IBM and were challenged to continue their quest to build a chess-playing machine that could defeat the human world champion. Deep Blue won its first chess game against a world champion in 1996, when it defeated Garry Kasparov in the first game of a match. At the end of the match, however, Kasparov had defeated Deep Blue by a score of 4 to 2. Deep Blue was then improved. In 1997 it played Kasparov again and became the first computer system to defeat a reigning world champion in a standard chess match. Kasparov never accepted the defeat. Afterward he said that he sometimes had observed deep intelligence and creativity in the machine's moves, suggesting that human chess players had helped the machine. IBM always denied that any cheating had taken place, noting that the rules had allowed the developers of Deep Blue to "tune" the machine's program between games and that they had made extensive use of that opportunity. The rematch demanded by Kasparov never took place. IBM ended up discontinuing Deep Blue. Whether Deep Blue was indeed better than the reigning chess champion is, of course, irrelevant. Deep Blue and many other chess computers in existence today probably can beat 99,999999 percent of humans in a game that was once thought to require strong artificial intelligence to be mastered. Deep Blue played chess in a way that was not, in many respects, the same way a human plays, but it can't be denied that Deep Blue had a field-specific mind for chess.

There are games more complex than chess. Go, which originated in China more than 2,500 years ago, is played on a board with 19 × 19 positions (figure 10.2). Players take turns placing black and white stones in order to control territories. Once played, the stones are never moved. There are hundreds of possible moves at each turn. Go is very difficult for computers because of the high branching factor and the depth of the search tree. Until recently, no Go-playing program could beat a reasonably good human player, and it was believed that playing Go well would remain beyond the capabilities of computers for a long time. However, in 2016 a team of researchers from a company called Google Deepmind used neural networks trained using deep learning (Silver et al. 2016) to create a program, called

Figure 10.2
The final position of a game of Go, with the territories defined.

AlphaGo, that played the game at the highest level. In January of 2016, AlphaGo defeated the European champion, Fan Hui. Two months later, it defeated the man who was considered the best player in the world, Lee Sedol.

When they play chess, Go, or similar games, computers typically perform a search, trying many moves and evaluating the millions of positions that may result from them. Evaluating a position in chess or in Go is difficult, but the number of pieces on the board and their positions can be used to obtain a good estimate of the position value, particularly if large databases of games are available. Although we don't know exactly how humans play such games, we know that computers use an approach based mostly on brute-force search, trying many possible moves before choosing one. Checkers has been completely solved by computers—that is, all possible combinations of moves have been tried and have been registered in a database, so that the best move in each situation is known to a program that has access to the database. This implies no human player without access to the database can ever beat such a computer. (A human with access to the database would also play the game perfectly, but no human's memory would be able to store such a huge amount of data.)

Games more complex than checkers, such as chess and Go, probably will never be solved exactly, because the number of possible games is simply too large. Although the exact number of possible games of chess isn't known, it is estimated to exceed 10^{120}, a number vastly larger than the number of atoms in the universe. The number of possible Go games is probably larger than 10^{800}. Therefore, it cannot be argued that Deep Blue or AlphaGo play by brute force (that is, by knowing all the possible positions). Deep Blue definitely had a mind for chess and AlphaGo a mind for Go, albeit a mind different from the minds of human players.

A somewhat less obvious application of artificial intelligence is used by many people almost every day. As of this writing, the World Wide Web is composed of tens of billions of Web pages. Google, the best-known and most widely used search engine, gives everyone access to the information contained in about 50 billion of those pages. By using a technique called *indexing*, in which the pages containing a certain term are listed together in an index organized a bit like a phone book, Google can retrieve, in a fraction of a second, the pages related to any set of topics chosen by a user. Furthermore, it is able to list the most relevant pages first. It does so by applying a number of clever methods that improved on the original idea of its founders (Brin and Page 1998), the page-rank algorithm.

The page-rank algorithm, which is at the origin of Google's success, views the Web as a huge, interconnected set of pages, and deems most likely to be interesting the pages visited more often if a user was randomly traveling this network by following the hyperlinks that connect the pages. Pages and domains with many links pointing to them are, therefore, deemed more interesting. This is why we are more likely to retrieve a page from CNN or BBC than one from some more obscure news source. Many more pages point to CNN (as a source of news or for some other reason) than to other, less-well-known pages. The page-rank algorithm has been, over the years, complemented by many other techniques that contribute to a good ranking of the pages retrieved as results of a query.

The apparently simple operation of retrieving a set of pages relevant to some query is the result of the application of very sophisticated technology, developed over the years by engineers and researchers in computer science. Many of us are now so accustomed to going to Google and retrieving relevant information on any obscure topic that we may forget that such an operation would have been very slow, if possible at all, only twenty years ago.

We usually don't think of Google as having a mind of its own, not even a field-specific mind. Let us, however, go back in time to 1985. Only a few years earlier, the TCP/IP protocol, which enabled computers to talk with other computers over long distances, had been developed and deployed. In 1985 one could use computer-to-computer networks to send and receive messages (mainly email and files). But only a few thousand people used it for that purpose, and no one predicted that, only a few decades later, there would be billions of computers and mobile devices interconnected as they are now. However, the essential technology already existed. Computers could already talk to one another. The only things that were needed to have the Internet we have today were more sophisticated algorithms and programs, faster computers, and faster interconnection networks.

Now imagine that you are back in 1985 and that someone tells you that, thirty years in the future, a computer system will exist that will be able to recover, in a fraction of a second, all the relevant information in the world about a given topic, chosen by you. You are told that such a system will go through all the information in the Internet (the roughly 10^{24} bytes stored in all the interconnected computers in the world), understand enough of it to select the information that is relevant to your request, and present the answer to you in a fraction of a second. To put things in perspective, it is useful to remember that a typical book consists of about half a million bytes, that the Library of Congress stores about 150 million books, and that therefore the information available from the Internet corresponds to about 10 billion times the information stored in the Library of Congress. (To be fair, the large majority of the data stored in the Internet corresponds to videos and photos, which are only partially covered by search engines, since these engines index videos and photos mostly on the bases of keywords and tags, not on the basis of the contents. However, that will soon change as computers get better at processing images, and it isn't really relevant to the question that will be posed next.)

Now, imagine that, back in 1985, you are asked: Is such a system intelligent? Does it have a mind of its own? Do you need intelligence to search this huge amount of information and, from it, retrieve instantly the relevant pieces of data? Does the system need to understand enough of what is written in the documents to retrieve relevant pieces of information and to discard the irrelevant ones?

My guess is that, back in 1985, you would have answered that such a system must be intelligent and must have a mind of its own, at least in some limited way. After all, to answer the complex queries it accepts, it must have some model of the world and it must understand, at least partially, what has been asked.

If you didn't know, at the time, how queries would be posed to a system such as the one we have now, you might have imagined some sort of written or spoken dialogue. For instance, a user would type some question (perhaps "What is the size of the Library of Congress?") and would get an answer something like "The Library of Congress is the largest library in the world, with more than 158 million items on approximately 838 miles of bookshelves." If you try that today, you probably will get back a document that includes exactly the information you were looking for, but it probably will be mixed with some additional information that may or may not be relevant to you. You probably are so accustomed to using a search engine that you don't even remotely consider that it may have a mind, but in 1985 you may have speculated that it did. There are some excellent reasons for this duality of criteria.

One of the reasons is that search engines are still far from having a perfect interface with the user and from retrieving exactly what the user wants. We talk to them mostly by means of text (even though we now can use a somewhat less than perfect voice interface), and, instead of replying by writing an explicit answer, they provide us with a list of documents in which the answers to our queries can be found. The technologies needed to transform such answers into answers in plain English are probably within reach. Indeed, such capabilities have been demonstrated by other systems. IBM's Watson is one such system.

Watson uses advanced algorithms for the processing of natural language, automated reasoning, and machine learning to answer questions in specific domains. Using information from Wikipedia and other knowledge sources, including encyclopedias, dictionaries, articles, and books, Watson was able to compete with human champions on the TV game show *Jeopardy*, beating the top human competitors a number of times. (In the game, what ordinarily would be considered questions are presented as answers, and players must respond with the relevant questions. For the purposes of this discussion, however, the former will be referred to as questions and the latter as answers.)

Watching videos of the *Jeopardy* competition (which are available in YouTube) is a humbling and sobering experience. A computer, using complex but well-understood technologies, interprets difficult questions posed in natural language and answers them faster and more accurately than skilled human players. Watson's computer-synthesized voice answers difficult questions with astounding precision, exhibiting knowledge of culture and of historical facts that very few people on Earth (if any) possess.

Yet skeptics were quick to point out that Watson didn't understand the questions, didn't understand the answers, and certainly didn't understand that it had won a game. One of the most outspoken skeptics was John Searle, whose arguments were based on the same ideas that support the Chinese Room experiment (discussed in chapter 5). All Watson does, Searle argued in a paper published in 2011, is manipulate symbols according to a specific set of rules specified by the program. Symbol manipulation, he argued, isn't the same thing as understanding; therefore, Watson doesn't understand any of the questions or the answers, which are meaningless symbols to the program. The system cannot understand the questions, because it has no way to go from symbols to meaning. You may ask "Then how does the brain do that?" Searle's answer is that a human brain, by simply operating, causes consciousness, understanding, and all that comes with consciousness and understanding.

I do not agree with Searle and the other skeptics when they say that Watson didn't understand the questions or the answers it provided. I believe that understanding is not a magical process, a light either on or off. Understanding is a gradual thing. It results from creating an internal model of the world that is relevant to the problem at hand. A system that has no understanding cannot answer, as accurately as Watson did, the complex questions posed in a game of *Jeopardy*. The model of the world in Watson's mind was enough to enable it to answer, accurately, a large majority of the questions. Some it didn't understand at all; others it understood partially; many—probably most—it understood completely.

The question, then, is not whether Watson is intelligent. Certainly it is intelligent enough to answer many complex questions, and if it can do so it must have a mind of his own—a field-specific mind for the particular game called *Jeopardy*. Did Watson understand that it had won the game? It probably did, since its behavior was directed at winning the game.

Was Watson happy for having won? Was it conscious of having won? Probably not, since consciousness of having won would have required a deeper understanding of the game, of the world, and of Watson's role in the world. As far as I know, Watson doesn't have internal models for those concepts, and doesn't have a sense of self, because internal models and a sense of self were not included in its design.

Watson has been used in medical diagnosis and in oil prospecting, domains in which its ability to answer questions is useful. Its ability to answer questions posed in natural language, coupled with its evidence-based learning capabilities, enable it to work as a clinical decision-support system. It can be used to aid physicians in finding the best treatment for a patient. Once a physician has posed a query to the system describing symptoms and other factors, Watson parses the input, looks for the most important pieces of information, then finds facts relevant to the problem at hand, including sources of data and the patient's medical and hereditary history. In oil prospecting, Watson can be an interface between geologists and the numerical algorithms that perform the complex signal-processing tasks used to determine the structure of potential underground oil fields.

Deep Blue, AlphaGo, Google, and Watson are examples of systems designed, in top-down fashion, to exhibit intelligence in specific fields. Deep Blue and AlphaGo excel at complex board games, Google retrieves information from the Web in response to a query, and Watson answers questions in specific domains in natural language. Other systems likely to be designed and deployed in the next decade are self-driving vehicles with a model of the world good enough to steer unassisted through roads and streets, intelligent personal assistants that will help us manage our daily tasks, and sales agents able to talk to customers in plain English. All these abilities have, so far, been unique to human beings, but they will certainly be mastered by machines sometime in the next decade.

Systems designed with the purpose of solving specific problems, such as those mentioned above, are likely to have field-specific minds sophisticated enough to enable them to become partners in conversations in their specific fields, but aren't likely to have the properties that would lead to general-purpose minds. But it is possible that, if many field-specific minds are put together, more general systems will become available. The next two decades will probably see the emergence of computerized personal assistants that can master such tasks as answering questions, retrieving information from the Web, taking care of personal agendas, scheduling meetings, and taking care of the shopping and provisioning of a house. The technologies required for those tasks will certainly be available in the next twenty years, and there will be a demand for such personal assistants (perhaps to be made available through an interface on a cell phone or a laptop computer).

Such systems will have more general minds, but they will still be limited to a small number of field-specific environments. To perform effectively they will have to be familiar with many specifics of human life, and to respond efficiently to a variety of requests they will have to maintain memory of past events.

I anticipate that we will view such a system as having a mind of its own, perhaps a restricted and somewhat simplified mind. We may still not consider such a system to have consciousness, personality, emotions, and desires. However, it is very likely that those characteristics are largely in the eye of the beholder. Even simple systems that mimic animals' behaviors and emotions can inspire emotions and perception of consciousness, as becomes apparent when one reads the description of the Mark III Beast in Terrel Miedaner's 1978 novel *The Soul of Anna Klane* (reprinted in Hofstadter and Dennett's book *The Mind's I*.) If Watson or Google were to be modified to interact more deeply with the emotions of its users, it is very possible that our perceptions about their personalities, emotions, and desires would change profoundly.

Kevin Kelly, in his insightful book *What Technology Wants*, uses the term *the technium* to designate the vast interconnected network of computers, programs, information, devices, tools, cities, vehicles, and other technological implements already present in the world. Kelly argues that the technium is already an assembly of billions of minds that we don't recognize as such simply because we have a "chauvinistic bias" and refuse to accept as mindful anything other than a human being.

To Kelly, the technium is the inevitable next step in the evolution of complexity, a process that has been taking place on Earth for more than 4 billion years and in the universe for much longer, since before Earth existed. In his view, the technium is simultaneously an autonomous entity, with its own agenda, and the future of evolution. It is a new way to develop more and more powerful minds, in the process creating more complexity and opening more possibilities. It is indeed possible that we are blind, or chauvinistic, or both, when we fail to recognize the Internet as an ever-present mindful entity, with its billions of eyes, ears, and brains. I don't know whether we will ever change our view of the synthetic minds we created and developed.

Neuromorphic Intelligent Systems

There are, however, other ways to design intelligent systems. One way is by drawing more inspiration from the human brain, the system that supports the only general mind we know. Intelligent systems designed in such a way will raise much more complex questions about what a mind is, and about the difference between synthetic minds and natural minds. If we design systems based on the same principles that are at work on human minds, we probably will then be more willing to accept that they have minds of their own and are conscious. After all, if one wants to create an intelligent and conscious system, the most obvious way is to copy the one system we know to be intelligent and conscious: the human mind. There are two ways to make such a copy.

One way would be to directly copy a working human mind. This could be done by copying, detail by detail, a human brain, the physical system that supports the human mind. Copying all the details of a working brain would be hard work, but the result, if successful, would certainly deliver what was desired. After all, if the resulting system were a complete and fully operational working copy of the original, and its behavior were indistinguishable from the original, certainly it would be intelligent. Because such a mind would be a piece-by-piece copy of an existing biological mind, it would behave as a natural mind behaves, even if it was supported by a digital computer. This mind design process, usually called mind uploading or whole-brain emulation, is the topic of the next section.

Another way to design a human-like mind is to reproduce, rather than the details of a working brain, the mechanisms that led to the creation of such a brain. This approach may be somewhat less challenging, and may be more accessible with technology likely to be available in the near future. The idea here is to reproduce in a computer not the details of a working brain, but the principles and the mechanisms that lead to the creation of the structures in a human brain, and to use these principles to create synthetic brains.

In chapter 8 I described the two main factors that are at work when a brain assembles itself: genetic encoding and brain plasticity. With the development of our knowledge of those two mechanisms, it may one day be possible to reproduce, in a program, the way a brain organizes itself.

Many researchers believe that a complex system such as a brain results, not from a very complex recipe with precisely tuned parameters and detailed connectivity instructions, but from a general mechanism that creates complexity from simple local interactions, subject to some very general rules that work within a wide range of parameters (Kelso 1997). Significant evidence obtained from firing patterns in the brain, and particularly in the cortex (Beggs and Plenz 2003), supports the idea that the brain organizes itself to be in some sort of critical state such that exactly the right amount of firing optimizes information transmission, while avoiding both activity decay and runaway network excitation.

Information about general mathematical principles of self-organization and detailed knowledge of the biological and biochemical mechanisms that control brain development may eventually enable us to simulate, in a computer, the processes involved in brain development. The simulations will have to be extensively compared against data obtained from both simple and complex organisms. If successful, the simulation should derive connectivity structures and patterns of cell activity that are, in all observable aspects, equivalent to those observed in real brains. If the physical and mathematical principles are sound, and if enough knowledge is available with which to tune the parameters, it is reasonable to expect that the resulting self-organized brain will behave, in most if not all respects, like a real brain. Such a simulation need not be performed at the level of neurons, since higher-level structures may be involved in the process of brain organization. In fact, neurons may not even need to be involved in the simulation at all, because they may not represent the right level of abstraction.

Of course, no human-like brain will develop normally if it doesn't receive a complete and varied set of stimuli. Those stimuli, which will have to enter the system through simulated senses, will play an essential part in the development of the right brain structures. The whole process of designing a brain by this route will take time, since the mechanisms that lead to brain organization take years to develop. In this process, interaction with the real world is one limiting factor. There is, in principle, no reason why the whole process cannot run many times faster than what occurs in a real brain. In fact, it is likely that the first approaches will use this possibility, making us able to simulate, in only a few days, development processes that take months or years to occur. However, as the simulation becomes more precise and more complex, there will be a need to provide the simulated brain with realistic input patterns, something that will require real-time interaction with the physical world. If the aim is to obtain, by this route, a working simulation of a brain of an infant monkey, the simulation will have to interact with other monkeys, real or simulated. If the aim is to simulate the full development of a human brain, interactions with humans and other agents will be required. This means that the complete simulation will have to be run, mainly, in real time, and it will take a few years to develop the simulation of a toddler and about twenty years to develop the simulation of a young adult.

Many complex questions are raised by the possibility that a human brain can be "developed" or "grown" in this way from first principles, experimental data, and real-world stimuli. If such an approach eventually leads to a system that thinks, understands speech, perceives the world, and interacts with humans, then it probably will be able to pass, with flying colors, any sort of unrestricted Turing Test. Even if interaction is always performed through a simulated world (one that never existed physically), this fact will not make the existence of an intelligent and conscious system less real or its emotions less genuine.

Such a mind would be synthetic, but its workings would be inspired by the workings of a natural mind, so it would be somewhat of a hybrid of a natural and a synthetic mind (see figure 10.1). If such a system is developed, we will have to relate to the first mind that is, in some respects, similar to our own, but that doesn't have a body and wasn't born of a human. His or her brain will perform computations similar to those performed by human beings, he or she will have feelings and will have the memories of a life lived in a world that exists only inside the memory of a computer. His or her feelings will presumably be no less real than ours, the emotions will be as vivid as ours, and the desire to live will be, probably, as strong as any of ours. For the first time, we would have created a synthetic mind modeled after a natural human mind.

Whole-Brain Emulation

As referred above, there is another way to create a human-like mind, and it consists in copying directly the working brain of a living human. Such a process would yield a digital mind that would also be a natural mind, because it would work exactly in the same way as a biological human mind. It would, however, be supported by a brain emulator running in a computer, rather than by the activity of a biological brain. This is, by far, the most ambitious and technologically challenging method, because it would require technology that doesn't yet exist and may indeed never exist. It also raises the most complex set of questions, because it changes completely the paradigm of what intelligence is and what it means to be human.

To create a digital mind from an existing human mind, it is necessary to reverse engineer a working brain in great detail. Conceptually there are many ways that might be done, but all of them require very detailed information about the structure and the patterns of neural activity of a specific human brain—information reliable and precise enough to make it possible to emulate the behavior of that brain in a computer. This approach, known as *whole-brain emulation* (WBE) or *mind uploading,* is based on the idea that complete emulation of a brain based on detailed information is possible. That whole-brain emulation will one day be possible has been proposed by a number of authors, including Ray Kurzweil and Nick Bostrom, but only recently have we begun to understand what it will take to attempt such a feat.

In theory, mind uploading might be accomplished by a number of methods, but two methods at different ends of the spectrum deserve special mention. One would entail copying neural structures wholesale; the other would entail gradual replacement of neurons. In the first method, mind uploading would be achieved by scanning and identifying the relevant structures in a brain, then copying that information and storing it in a computer system, which would use it to perform brain emulation. The second method would be more gradual. The original brain would keep operating, but its computational support would be progressively replaced.

The first approach, wholesale copying of a working brain, requires mainly enormous improvements in existing technologies coupled with the development of some new ones. In chapter 9 I discussed some techniques that could be used to identify detailed structures in human brains; in chapter 8 I discussed techniques that could be used to simulate sections of brain tissue if given detailed models of the neurons, their interconnectivity patterns, and other relevant biochemical information. Existing knowledge of the structure of brain tissue can, in principle, be used to simulate a human brain very accurately (even if slowly) if given a detailed description of each neuron and of how it interacts with other neurons. The challenge lies in the fact that no existing technology is sufficiently advanced to obtain the detailed structural information required to supply a brain emulator with the required level of detail.

At present, a number of sophisticated imaging methods can be used to retrieve information about brain structures and neuron activity. In chapter 9, we considered a number of techniques, among them PET and MRI, that can be used to obtain functional information about how the brain is wired and about the activity levels of the neurons involved in specific tasks. However, even the most powerful techniques in use today can only obtain, non-invasively, information about structures that are no smaller than a cubic millimeter, which is the size of the smallest voxel they can resolve. Within a cubic millimeter of cortex there are between 50,000 and 100,000 neurons, each with hundreds or thousands of synapses that connect with other neurons, for a total number of synapses that is on the order of a billion per cubic millimeter, each synapse ranging in size from 20 to 200 nanometers.

Therefore, present-day imaging techniques cannot be used to obtain detailed information about synapse location and neuron connectivity in live brains. The actual resolution obtained using MRI varies with the specific technique used, but resolutions as high as a billion voxels per cubic millimeter have been reported, which corresponds to voxels with 1 μm on the side (Glover and Mansfield 2002). There are, however, many obstacles that have to be surmounted to achieve such high resolution, and it is not clear whether such techniques could ever be used to image large sections of a working brain.

To resolve individual synapses would require microscopy techniques that can now be used to image slices of brain tissue with unprecedented detail. These techniques destroy the tissue and can be used only in dead brains. In chapter 7 we saw how the brain of the worm *C. elegans* was painstakingly sliced and photographed, using electron microscopy, to obtain the complete wiring diagram of its 302 neurons, in an effort that took 12 years to complete. If such an effort and methodology were to be applied to a human brain, it would require a time commensurate with the age of the universe, if we take into consideration only the scaling up implied by the much larger number of neurons and synapses.

However, technology has evolved, and it may one day become possible to apply a similar process effectively to a whole human brain. The process done mostly by hand in the *C. elegans* study can now be almost completely automated, with robots and computers doing all the brain slicing, electron microscopy, and three-dimensional reconstruction of neurons. Small volumes of neocortex have already been reconstructed in great detail, as was noted in chapter 9. With the right technology, such a process could conceptually be applied to a whole human brain, obtaining a very faithful copy of the original organ (which would, regrettably, be destroyed in the process). The level of detail gathered in this way would probably not be enough to emulate accurately the brain that was sliced and imaged, since significant uncertainty in a number of parameters influencing neuron behavior might remain, even at this level of resolution. However, it might be possible to fill in the missing information by combining data about the real-time behavior of the brain (obtained, of course, while it was still operating). Optogenetics, a technology to which I alluded briefly in chapter 9, could be used to provide very useful information about detailed brain behavior.

It is clear that existing technologies cannot easily be used to obtain information about working brains with a level of detail sufficient to feed a brain emulator that can generate sufficiently accurate output. Obtaining such detailed information from a working brain is clearly beyond the reach of existing technologies. On the other hand, structural information, obtained using microscopy, can be very detailed and could be used, at least in principle, to feed a brain emulator, even though much information would have to be obtained by indirect means. Limitations of state-of-the-art techniques still restrict the applicability of these techniques to small volumes, on the order of a fraction of a cubic millimeter, but we can expect that, with improved algorithms and better microscopy technologies, the whole brain of a mouse could be reverse engineered in great detail within a few decades.

However, it isn't clear that structural information obtained using existing microscopy techniques, such as serial block-face electron microscopy, has enough detail to feed an emulator that would faithfully reproduce the original system. The difficulty lies in the fact that the behavior of the system is likely to be sensitive to a number of parameters that are hard to estimate accurately from purely structural information, such as properties of membranes and synapses, and from chemical information present in the environment outside the neurons.

Ideally, detailed structural information would be combined with *in vivo* functional information. Such a combination of information might make it possible to tune the emulator parameters in order to reproduce, *in silico*, the behavior of the original system. Although many methods for combining these two types of information are under development (Ragan et al. 2012), combining functional information from a live brain and detailed structural information obtained by microscopy of a sliced brain remains a challenge. It is therefore not yet possible to use the significant amount of knowledge about the behavior of individual neurons and groups of neurons to construct a complete model of a brain, because the requisite information about neuron structure, properties, and connectivity is not yet available.

There is, in fact, a huge gap in our understanding of brain processes. On the one hand, significant amounts of information at the macro level are being collected and organized by many research projects. That information can be used to find out which regions of the brain are involved in specific processes, but cannot be used to understand in detail the behavior of single neurons or of small groups of neurons. It may, however, provide some information about the way the brain organizes itself at the level of macrostructures containing millions of neurons. On the other hand, researchers understand in significant detail many small-scale processes and structures that occur in individual neurons, including synapse plasticity and membrane firing behavior.

There is no obvious way to bridge this huge gap. Because of physical limitations, it isn't likely that existing techniques, by themselves, can be used to determine all the required information, without which there is no way to build an accurate brain emulator. Nanotechnologies and other techniques may be usable to bridge the gap, but they will not be used in healthy human brains before the technology is mature enough to guarantee results; even then, they will raise significant moral questions. Ongoing projects that use animal models, including OpenWorm, the Open Connectome Project, and the Human Brain Project, will shed some light on the feasibility of different approaches, but finding an exact way to proceed will remain the focus of many projects in coming decades.

I cannot tell you, because I don't know, how this challenge will be met. Many teams of researchers, all over the world, are working on techniques aimed at better understanding the brain. No single technique or combination of techniques has so far delivered on the promise of being able to obtain the detailed information that could be used to perform whole-brain emulation. One may even be tempted to argue that, owing to physical limitations and the complexity of the challenges involved, we will never have information detailed enough to enable us to build a faithful emulator.

In order for mind uploading to become possible, a number of technologies would have to be enormously developed, including advanced microscopy, computational techniques for three-dimensional reconstruction of brain tissue, and multi-level neural modeling and simulation. Enormous advances in processing power would also be necessary. A workshop on whole-brain emulation held at the Future of Humanity Institute in Oxford in 2007 produced a "road map" (Sandberg and Bostrom 2008) that discusses in detail the technologies required and the appropriate level of emulation. In theory, brain emulation could be performed at a variety of levels, or scales, including quantum, molecular, synaptic, neuronal, or neuron population. Depending on the exact mechanisms used by brains, emulation might have to be performed at a more or less detailed level. The level of the emulation has a significant effect on the computational resources required to perform it. Quantum or molecular-level emulation will probably be forever prohibitive (barring extraordinary advances in quantum computing). Higher-level emulation is computationally more efficient but may miss some effects that are essential in brain behavior.

The road map also includes a number of milestones that could be used to measure how far in the future whole-brain emulation is. It concludes by estimating, perhaps somewhat optimistically, that insufficient computational power is the most serious impediment to achieving whole-brain emulation by the middle of the twenty-first century. That conclusion, however, rests on the assumption that all the supporting technologies would have advanced enough by that time, something that is far from widely agreed upon.

One important argument presented in the road map is that whole-brain emulation doesn't really require the creation of yet unknown technologies, but only scaling up (and vast improvement) of already-existing ones. After all, we already have the technologies to reverse engineer small blocks of brain tissue, the models that enable us to simulate neural tissue, and the simulators that use these models to perform simulations on a relatively small scale. All that is required, now, is to improve these technologies, so that we can apply them first to simple invertebrates, like *C. elegans*, then to larger and larger animals, including progressively more complex mammals, until we are finally able to emulate a complete human brain. None of these developments requires conceptually new technologies, although all of them require enormous advances in existing technologies.

The second approach to mind uploading, gradual replacement of neurons in a working brain, requires technologies even more complex and distant than the ones required to perform a wholesale copy. One such technology, which is still in its infancy but which may come to play an important role in the future, is nanotechnology.

In his influential 1986 book *Engines of Creation*, Eric Drexler popularized the idea that molecular-size machines could be assembled in large numbers and used to perform all sorts of tasks that are now impossible. The idea may have originated with Richard Feynman's lecture "There's Plenty of Room at the Bottom," given in 1959 at a meeting of the American Physical Society. In that lecture, Feynman addressed the possibility of direct manipulation of individual atoms to create materials and machines with unprecedented characteristics.

The ideas of the pioneers mentioned above have been developed by the scientific community in many different directions. Some researchers have proposed the development of techniques for constructing nanodevices that would link to the nervous system and obtain detailed information about it. Swarms of nanomachines could then be used to detect nerve impulses and send the information to a computer. Ultimately, nanomachines could be used to identify specific neurons and, progressively, replace them with arrays of other nanomachines, which could then use the local environmental conditions, within the operating brain, to adapt their behavior. These nanomachines would have to be powered by energy obtained from electromagnetic, sonic, chemical, or biological sources, and proposals for these different approaches have already been made.

Nanotechnology has seen many developments in recent decades. Many types of nanoparticles with diameters of only a few nanometers have been created in the laboratory, as have carbon nanotubes with diameters as small as one nanometer, nano-rotors, and nano-cantilevers. Chips that sort and manipulate individual blood cells have been developed and will be deployed soon, making it possible to identify and remove cancer cells from blood (Laureyn et al. 2007). However, the technologies necessary to design swarms of autonomous nanorobots, which could perform the complex tasks required to instrument a living brain or to replace specific neurons, do not yet exist and will probably take many decades, or even centuries, to develop.

We are still a long way from the world of Neal Stephenson's 1995 novel *The Diamond Age*, in which swarms of nanorobots are used to interconnect human brains into a collective mind, but there is no obvious reason why such a technology should be forever impossible.

One of the most interesting thought experiments related to this question was proposed by Arnold Zuboff in "The Story of a Brain," included in the book *The Mind's I* (Hofstadter and Dennett 2006). Zuboff imagines an experiment in which the brain of a man who had died of a disease that hadn't affected his brain is put on a vat and fed with stimuli that mimicked those he would have received had he been alive. As the experiment evolved and more teams got involved, the brain was split several times into ever smaller pieces and eventually into individual neurons, all of which were connected by high-speed communication equipment that made sure the right stimuli reached every neuron. In the end, all that remained of the original brain was a vast network of electronic equipment, providing inputs to neurons that had nothing to do with the original brain and that could be, themselves, simulated. And yet, the reader is led to conclude, the original personality of the person whose brain was put into the vat was left untouched, his lifetime experiences unblemished, his feelings unmodified.

No matter which approach is used, if an emulation of a human brain can, one day, be run in a computer, we will be forced to face a number of very difficult questions. What makes us human is the processing of information that enables us to sense the world, to think, to feel, and to express our feelings. The physical system that supports this information-processing ability can be based on either biological or electronic digital devices. It isn't difficult to imagine that one small part of the brain, perhaps the optic nerve, can be replaced without affecting the subjective experience of the brain's owner. It is more difficult to imagine that the whole brain can be replaced by a digital computer running very detailed and precise emulation without affecting the subjective experience of the owner.

It is possible that someday a supercomputer will be able to run a complete emulation of a person's brain. For the emulation to be realistic, appropriate inputs will have to be fed into the sensory systems, including the vision pathways, the auditory nerves, and perhaps the neural pathways that convey other senses. In the same way, outputs of the brain that activate muscles should be translated into appropriate actions. It isn't likely that a faithful emulation of a complete brain can be accomplished without these components, because realistic stimuli (without which no brain can work properly) would be missing.

Greg Egan, in his 2010 book *Zendegi*, imagines a technology, not too far in the future, by which the brain behaviors of a terminally ill man, Martin, are scanned using MRI and used to create a synthetic agent that was expected to be able to interact with his son through a game-playing platform. In the end, the project fails because the synthetic agent created—the virtual Martin—doesn't have enough of Martin's personality to be able to replace him as a father to his son in the virtual environment of the game. The virtual Martin created from fragments of Martin's brain behavior is too far from human, and the researchers recognize that uploading the mind of a person into virtual reality is not yet possible. Despite the negative conclusion, the idea raises interesting questions: When will we be sure the copied behaviors are good enough? How do you test whether a virtual copy of a human being is good enough to replace, even in some limited domain, the original? What do you do with copies that aren't good enough?

It is safe to say no one will be able to understand in all its details a full-blown emulation of a working brain, if one ever comes into existence. Any decisions to be made regarding the future of such a project will have to take into consideration that the information processing taking place in such an emulation is the same as the information processing that goes on in a living brain, with all the moral and philosophical consequences that implies.

I am not underestimating the technical challenges involved in such a project. We are not talking about simulating a small subsystem of the brain, but about emulating a full brain, with its complete load of memories, experiences, reactions, and emotions. We are very far from accomplishing such a feat, and it is not likely that anyone alive today will live to see it. However, that isn't the same thing as saying that it will never be possible. In fact, I believe that it will one day be accomplished.

The possibility that such a technology may one day exist, many years or even centuries in the future, should make us aware of its philosophical, social, and economic consequences. If a brain emulator can ever be built, then it is in the nature of technology that many more will soon be built.

We are still a long way from having any of the technologies required to perform whole-brain emulation. I have, however, a number of reasons to believe that we will be able to overcome this challenge. One of these reasons is simply the intuition that technological developments in this area will continue to evolve at an ever-increasing speed. The exponential nature of technology tends to surprise us. Once some technology takes off, the results are almost always surprising. The same will certainly be true for the many technologies now being used to understand the brain, many of them developed in the past few decades. Though it may be true that no single technology will suffice, by itself, to derive all the information required to reconstruct a working brain, it may be possible that some combination of technologies will be powerful enough.

Other reasons to believe that we will, one day, be able to emulate whole brains are related to the enormous economic value of such a technology. Some economists believe that world GDP, which now doubles every 15 years or so, could double every few weeks in the near future (Hanson 2008). A large part of these gains would be created by the creation of digital minds in various forms.

The ability to emulate a working brain accurately will create such radical changes in our perspective of the world and in our societies that all other challenges we are currently facing will become, in a way, secondary. It is therefore reasonable to believe that more and more resources will be committed to the development of such a technology, making it more likely that, eventually, all existing barriers will be overcome.

The Riddle of Consciousness

Should complex and very general artificial minds come into existence, we will have to address in a clear way the important question of whether they are conscious.

We all seem to know very well what it means to be conscious. We become conscious when we wake up in the morning, remain conscious during waking hours, and lose consciousness again when we go to sleep at night. There is an uninterrupted flow of consciousness that, with the exception of sleeping periods, connects the person you are now with the person you were many years ago and even with the person you were at birth, although most people cannot track down their actual memories that far back. If pressed to describe how consciousness works, most people will fall into one of two main groups: the dualists and the monists.

Dualists believe there are two different realms that define us: the physical realm, well studied and understood by the laws of physics, and the non-physical realm, in which our selves exist. Our essence—our soul, if you want—exists in the non-physical realm, and it interacts with and controls our physical body through some yet-unexplained mechanism. Most religions, including Christianity, Islam, and Hinduism, are based on dualist theories. Buddhism is something of an exception, since it holds that human beings have no permanent self, although different schools of Buddhism, including the Tibetan one, have different ideas about exactly what characteristics of the self continue to exist after death. Religions do not have a monopoly on dualism—the popular New Age movement also involves theories about the non-physical powers of the mind.

The best-known dualist theory is attributable to René Descartes. In *Meditationes de Prima Philosophia* (1641), he proposes the idea (now known as Cartesian dualism) that the mind and the brain are two different things, and that, whereas the mind has no physical substance, the body, controlled by the brain, is physical and follows the laws of physics. Descartes believed that the mind and the body interacted by means of the pineal gland. Notwithstanding Descartes' influence on modern philosophy, there is no evidence whatsoever of the existence of a non-physical entity that is the seat of the soul, nor is there evidence that it interacts with the body through the small pineal gland.

Cartesian dualism was a response to Thomas Hobbes' materialist critique of the human person. Hobbes, whose mechanist view was discussed in chapter 2, argued that all of human experience comes from biological processes contained within the body. His views became prevalent in the scientific community in the nineteenth century and remain prevalent today. Indeed, today few scientists see themselves as dualists. If you have never thought about this yourself, a number of enlightening thought experiments can help you decide whether you are a dualist or a monist.

Dualists, for instance, believe that zombies can exist, at least conceptually. For the purposes of this discussion, a zombie behaves exactly as a normal person does but is not conscious. If you believe that a zombie could be created (for instance, by assembling, atom by atom, a human being copied from a dead person, or by some other method) and you believe that, even though it is not a person, its behavior would be, in all aspects, indistinguishable from the behavior of a real person, you probably are a dualist. Such a zombie would talk, walk, move, and behave just like a person with a mind and a soul, but would not be a conscious entity; it would merely be mimicking the behavior of a human.

To behave like a real person, a zombie would have to have emotions, self-awareness, and free will. It would have to be able to make decisions based on its own state of mind and its own history, and those decisions would have to be indistinguishable from those of a real human. Raymond Smullyan's story "An Unfortunate Dualist" (included in Smullyan 1980) illustrates the paradox that arises from believing in dualism. In the story, a dualist who wants to commit suicide but doesn't want to make his friends unhappy learns of a wonderful new drug that annihilates the soul but leaves the body working exactly as before. He decides to take the drug the very next morning. Unbeknownst to him, a friend who knows of his wishes injects him with the drug during the night. The next morning, the body wakes up without a soul, goes to the drugstore, takes the drug (again), and concludes angrily that the drug has no discernible effect.

We are forced to conclude that a creature that behaves exactly like a human being, exhibits the same range of emotions, and makes decisions in exactly the same way has to be conscious and have a mind of its own—or, alternatively, that no one is really conscious and that consciousness is an illusion.

A counter-argument based on another thought experiment was proposed by Donald Davidson (1987). Suppose you go hiking in a swamp and you are struck and killed by a lightning bolt. At the same time, nearby in the swamp, another lightning bolt spontaneously rearranges a bunch of different molecules in such a way that, entirely by chance, your body is recreated just as it was at the moment of your death. This being, whom Davidson calls Swampman, is structurally identical to you and will, presumably, behave exactly as you would have behaved had you not died. Davidson holds there would be, nevertheless, a significant difference, though the difference wouldn't be noticeable by an outside observer. Swampman would appear to recognize your friends; however, he wouldn't be able to recognize them, not having seen them before. Swampman would be no more than a zombie. This reasoning, which, to me, is absurd, is a direct consequence of a dualist view of the consciousness phenomenon, not much different from the view Descartes proposed many centuries ago.

Monists, on the other hand, don't believe in the existence of dual realities. They believe that the obvious absurdity of the concept of zombies shows that dualism is just plainly wrong. The term *monism* was first used by Christian Wolff, in the eighteenth century, to designate the position that everything is either mental (idealism) or physical (materialism) in order to address the difficulties present in dualist theories. (Here I will not discuss idealism, which is at significant odds with the way present-day science addresses and studies physical reality.)

Either implicitly or explicitly, present-day scientists are materialists at heart, believing that there is only one physical reality that generates every observable phenomenon, be it consciousness, free will, or the concept of self.

Even though you may reject dualism, you may still like to think there is an inner self constantly paying attention to some specific aspects of what is going on around you. You may be driving a car, with all the processing it requires, but your conscious mind may be somewhere else, perhaps thinking about your coming vacation or your impending deliverable. The inner self keeps a constant stream of attention, for every waking hour, and constructs an uninterrupted stream of consciousness. Everything works as if there is a small entity within each one of us that looks at everything our senses bring in but focuses only on some particular aspects of this input: the aspects that become the stream of consciousness. The problem with this view is that, as far as we know, there is no such entity that looks at the inputs, makes decision of its own free will, and acts on those decisions. There is no part of the brain that, if excised, keeps everything working the same but removes consciousness and free will. There isn't even any detectable pattern of activity that plays this role. What exists, in the brain, are patterns of activity that respond to specific inputs and generate sequences of planned actions.

Benjamin Libet's famous experiments (see Libet et al. 1983) have shown that our own perception of conscious decisions to act deceives us. Libet asked subjects to spontaneously and deliberately make a specific movement with one wrist. He then timed the exact moment of the action, the moment a change in brain activity patterns took place, and the moment the subjects reported having decided to move their wrist. The surprising result was not that the decision to act came before the action itself (by about 200 milliseconds), but that there was a change in brain activity in the motor cortex (the part of the cortex that controls movements) about 350 milliseconds before the decision to act was reported. The results of this experiment, repeated and confirmed in a number of different forms by other teams, suggest that the perceived moment of a conscious decision occurs a significant fraction of a second after the brain sets the decision in motion. So, instead of the conscious decision being the trigger of the action, the decision seems to be an after-the-fact justification for the movement. By itself this result would not prove there is no conscious center of decision that is in charge of all deliberate actions, but it certainly provides evidence that consciousness is deceiving in some aspects. It is consistent with this result to view consciousness as nothing more than the perception we have of our own selves and the way our actions change the environment we live in.

Many influential thinkers view the phenomenon of consciousness as an illusion, an attempt made by our brains to make sense of a stream of events perceived or caused by different parts of the brain acting independently (Nørretranders 1991). Dennett, in his book *Consciousness Explained* (1991), proposed the "multiple drafts" model for consciousness, which is based on this idea. In this model, there are, at any given time, various events occurring in different places in the brain. The brain itself is no more than a "bundle of semi-independent agencies" fashioned by millions of years of evolution. The effects of these events propagate to other parts of the brain and are assembled, after the fact, in one single, continuous story, which is the perceived stream of consciousness. In this model, the conscious I is not so much the autonomous agent that makes decisions and issues orders as it is a reporter that builds a coherent story from many independent parallel events. This model may actually be consistent, up to a point, with the existence of zombies, since this after the fact assembly of a serial account of events does not necessarily change that much the behavior of the agent, in normal circumstances.

This model is consistent with the view that consciousness may be an historically recent phenomenon, having appeared perhaps only a few thousand years ago. In his controversial 1976 book *The Origin of Consciousness in the Breakdown of the Bicameral Mind*, Julian Jaynes presents the argument that before 1000 BC humans were not conscious and acted mostly on the basis of their instincts, in accordance with the "inner voices" they heard, which were attributed to the gods. In Jaynes' theory, called *bicameralism*, the brain of a primitive human was clearly divided into two parts: one that voiced thoughts and one that listened and acted. The end of bicameralism, which may have been caused by strong societal stresses and the need to act more effectively toward long-term objectives, marked the beginning of introspective activities and human consciousness.

We may never know when and exactly how human consciousness appeared. However, understanding in more objective terms what consciousness is will become more and more important as technology advances. Determining whether a given act was performed consciously or unconsciously makes a big difference in our judgment of many things, including criminal acts.

The problem of consciousness is deeply intermixed with that other central conundrum of philosophy, free will. Are we entities that can control our own destiny, by making decisions and defining our course of action, or are we simply complex machines, following a path pre-determined by the laws of physics and the inputs we receive? If there is no conscious I that, from somewhere outside the brain, makes the decisions and controls the actions to be taken, how can we have free will?

The big obstacle to free will is determinism—the theory that there is, at any instant, exactly one physically possible future (van Inwagen 1983). The laws of classical physics specify the exact future evolution of any system, from a given starting state. If you know the exact initial conditions and the exact inputs to a system, it is possible, in principle, to predict its future evolution with absolute certainty. This leaves no space for free-willed systems, as long as they are purely physical systems.

Quantum physics and (to some extent) chaotic systems theory add some indetermination and randomness to the future evolution of a system, but still leave no place for free will. In fact, quantum mechanics adds true randomness to the question and changes the definition of determinism. Quantum mechanics states that there exist many possible futures, and that the exact future that unfolds is determined at the time of the collapse of the wave function. Before the collapse of the wave function, many possible futures may happen. When the wave function collapses and a particular particle goes into one state or another, a specific future is chosen. Depending on the interpretation of the theory, either one of the possible futures is chosen at random or (in the many-worlds interpretation of quantum mechanics) they all unfold simultaneously in parallel universes. In any case, there is no way this genuine randomness can be used by the soul, or the conscience, to control the future, unless it has a way to harness this randomness to choose the particular future it wants.

Despite this obvious difficulty, some people believe that the brain may be able to use quantum uncertainty to give human beings a unique ability to counter determinism and impose their own free will. In *The Emperor's New Mind*, Roger Penrose presents the argument that human consciousness is non-algorithmic, and that therefore the brain is not Turing equivalent. Penrose hypothesizes that quantum mechanics plays an essential role in human consciousness and that some as-yet-unknown mechanism working at the quantum-classical interface makes the brain more powerful than any Turing machine, at the same time creating the seat of consciousness and solving the dilemma of determinism versus free will. Ingenious as this argument may be, there is no evidence at all such a mechanism exists, and, in fact, people who believe in this explanation are, ultimately, undercover dualists.

Chaos theory is also viewed by some people as a loophole for free will. In chaotic systems, the future evolution of a system depends critically and very strongly on the exact initial conditions. In a chaotic system, even very minor differences in initial conditions lead to very different future evolutions of a system, making it in practice impossible to predict the future, because the initial conditions can never be known with the required level of precision. It is quite reasonable to think that the brain is, in many ways, a chaotic system. Very small differences in a visual input may lead to radically different perceptions and future behaviors. This characteristic of the brain has been used by many to argue against the possibility of whole-brain emulation and even as a possible mechanism for the existence of free will. However, none of these arguments carries any weight. Randomness, true or apparent, cannot be the source of free will and is not, in itself, an obstacle for any process that involves brain emulation. True, the emulated brain may have a behavior that diverges rapidly from the behavior of the real brain, either as a result of quantum phenomena or as a result of the slightly different initial conditions. But it remains true that both behaviors are reasonable and correspond to possible behaviors of the real brain. In practice, it is not possible to distinguish, either subjectively or objectively, which of these two behaviors is the real one.

Since quantum physics and chaos theory don't provide a loophole for free will, either one has to be a dualist or one has to find some way to reconcile free will and determinism—a difficult but not impossible task. In his thought-provoking book *Freedom Evolves*, Dennett (2003) tries to reconcile free will and determinism by arguing that, although in the strict physical sense our future is predetermined, we are still free in what matters, because evolution created in us an innate ability to make our own decisions, independent of any compulsions other than the laws of nature. To Dennett, free will is about freedom to make decisions, as opposed to an impossible freedom from the laws of physics.

There is a thought experiment that can be used to tell whether, at heart, you are a dualist, even if you don't consider yourself a dualist. In *Reasons and Persons*, Derek Parfit (1984) asks the reader to imagine a teletransporter, a machine that can, instantly and painlessly, make a destructive three-dimensional scan of yourself and then send the information to a sophisticated remote 3D assembler. The assembler, using entirely different atoms of the very same elements, can then re-create you in every detail. Would you use such a machine to travel, at the speed of light, to any point in the universe where the teletransporter can send you? If you are a true monist, and believe that everything is explained by physical reality, you will view such a machine simply as a very sophisticated and convenient mean of transport, much like the assembler-gates (A-gates) in Charles Stross' 2006 novel *Glasshouse*. On the other hand, if there are some dualist feelings in you, you will think twice about using such a device for travel, since the you emerging at the other end may be missing an important feature: your soul. In *Glasshouse*, A-gates use nanotechnology to recreate, atom by atom, any physical object, animate or inanimate, from a detailed description or from 3D scans of actual objects. An A-gate is a kind of a universal assembler that can build anything out of raw atoms from a given specification.

A-gates create the possibility of multiple working copies of a person—an even more puzzling phenomenon, which we will discuss in the next chapter. Teletransporters and A-gates don't exist, of course, and aren't likely to exist any time soon, but the questions they raise are likely to appear sooner than everyone expects, in the digital realm.

Where does all this leave us in regard to consciousness? Unless you are a hard-core dualist or you believe somehow in the non-algorithmic nature of consciousness (two stances that, in view of what we know, are hard to defend), you are forced to accept that zombies cannot exist. If an entity behaves like a conscious being, then it must have some kind of stream of consciousness, which in this case is an emergent property of sufficiently complex systems.

Conceptually, there could be many different forms of consciousness. Thomas Nagel's famous essay "What Is It Like to Be a Bat?" (1974) asks you to imagine how is the world perceived from the point of view of a bat. It doesn't help, of course, to imagine that you are a bat and that you fly around in the dark sensing your surroundings through radar, because a literate, English-speaking bat would definitely not be a typical bat. Neither does it help to imagine that, instead of eyes, you perceive the world through sonar, because we have no idea how bats perceive the world or how the world is represented in their minds. In fact, we don't even have an idea how the world is represented in the minds of other people, although we assume, in general, that the representation is probably similar to our own.

The key point of Nagel's question is whether there is such an experience as being a bat, in contrast with, say, the experience of being a mug. Nobody will really believe there is such a thing as being a mug, but the jury is still out on the bat question. Nagel concludes, as others have, that the problem of learning what consciousness is will never be solved, because it rests on the objective understanding of an entirely subjective experience, something that is a contradiction in terms. To be able to know what it is like to be a bat, you have to be a bat, and bats, by definition, cannot describe their experience to non-bats. To bridge the gap would require a bat that can describe its own experience using human language and human emotions. Such a bat would never be a real bat.

It seems we are back at square one, not one step closer to understanding what consciousness is. We must therefore fall back on some sort of objective black-box test, such as the Turing Test, if we don't know the exact mechanisms some system uses to process information. On the other hand, if we know that a certain system performs exactly the same computations as another system, even if it uses a different computational substrate, then, unless we are firm believers in dualism, we must concede that such a system is conscious and has free will.

Therefore, intelligent systems inspired by the behavior of the brain or copied from living brains will be, almost by definition, conscious. I am talking, of course, about full-fledged intelligent systems, with structures that have evolved in accordance to the same principles as human brains or structures copied, piece by piece, from functioning human brains.

Systems designed top-down, from entirely different principles, will raise a set of new questions. Will we ever consider a search engine such as Google or an expert system such as Watson conscious? Will its designers ever decide to make such a system more human by providing it with a smarter interface—one that remembers past interactions and has its own motivations and goals? Up to a point, some systems now in existence may already have such capabilities, but we clearly don't recognize these systems as conscious. Are they missing some crucial components (perhaps small ones), or is it our own anthropocentric stance that stops us from even considering that they may be conscious? I believe it is a combination of these two factors, together with the fact that their designers hesitate to raise this question—in view of the fear of hostile superhuman intelligences, it wouldn't be good for business. But sooner or later people will be confronted with the deep philosophical questions that will arise when truly intelligent systems are commonplace.

The next chapter discusses these thorny issues. The challenges that will result from the development of these technologies, and the questions they will raise, are so outlandish that the next chapter may feel completely out of touch with reality. If you find the ideas in this chapter outrageous, this may be a good time to stop reading, as the next chapters will make them seem rather mild.

11 Challenges and Promises

It is difficult to argue, with conviction, that digital minds will never exist, unless you are a firm believer in dualism. In fact, even though today we don't have the technologies necessary to create digital minds, it is difficult to believe that such technologies will never be developed. I will not try to predict when such technologies will be developed, but in principle there is no reason to doubt that some sort of sophisticated digital minds will become available during the twenty-first century.

Civil Rights

If a system is recognized as a digital mind, the thorny problem of civil rights will have to be addressed. We take for granted that existing legislation recognizes each human being as having a special status, the status of a person. Personhood, with the rights and responsibilities that come with it, is nowadays granted virtually to every living human in almost every society. However, the notions of personhood are not universal. Furthermore, the universal recognition that every human being, independently of race, sex, or status, has the right to personhood is historically recent. Slaves weren't recognized as persons before the abolition of slavery, a lengthy and slow process that took many centuries. In the United States, the nineteenth amendment to the Constitution, not ratified until 1920, granted women the right to vote in every state.

The concept of personhood varies somewhat with the legal system of the country. In the United States, as in many other countries, personhood is recognized by law because a person has rights and responsibilities. The person is the legal subject or substance of which the rights and responsibilities are attributes. A human being is called a "natural person," but the status of personhood can also be given to firms, associations, corporations, and consortia.

It seems reasonable to assume that discussions about the personhood status of digital minds will begin as soon as rights and responsibilities are attributed to them. Digital agents—software programs that take significant decisions on behalf of individuals and corporations—already exist and are responsible for many events affecting human lives. The most visible of them—automated trading agents, whose behavior is based on complex sets of rules and algorithms—affect the economy enormously and have been blamed for a number of economic crises and other troublesome events. Until now, however, the individuals or corporations that own those agents, rather than the agents themselves, have been held responsible for the consequences of their actions. This is simply a consequence of the fact that personhood has not, so far, been granted to any purely digital entity.

However, as digital agents and their behavior become more complex, it is natural to shift the responsibilities from the "owners" of these agents to the agents themselves. In fact, the concept of ownership of an intelligent agent may one day become a focus of dissension if digital minds are granted full personhood. The whole process will take a long time and will generate much discussion along the way, but eventually society will have to recognize digital minds as autonomous persons with their own motivations, goals, rights, and responsibilities. The question whether a digital mind has personhood will be unavoidable if the digital mind in question was created by means of mind uploading, a technology that may become feasible before the end of the century. Society will have to address the personhood status of such digital minds, providing them with some set of civil rights and responsibilities. I will call these entities simply *digital persons*, and will apply the term to any kind of digital mind that is granted some sort of personhood status. To simplify the writing, I will use the pronoun *it* to refer to digital persons, but this doesn't mean these persons are less human than actual humans, nor does it mean they do not have gender. Whether or not they will view themselves as male, female, or neither will depend on many factors, which I will not address here because so many variables are involved. I could as well have decided to address digital persons as *him*, *her*, or *it*, depending on the particular circumstances, but that would have been more awkward to write and to read.

The most central issue is probably related to the rights and responsibilities of digital persons. Suppose a program passes a full-fledged Turing Test by using advanced voice and visual interfaces to convince judges systematically, permanently, and repeatedly that it behaves like a human. To achieve that, a program would have to display the reasoning, memory, learning abilities, and even emotions that a human being would display under similar conditions. This program, a digital person, would have acquired much of its knowledge by means of techniques that enabled it to improve its behavior and to become more human-like as time passed. It is hard to imagine that a program could pass a full-fledged Turing Test without having some form of consciousness, but for now we can assume that it could simply be a very, very human-like zombie.

Digital persons might play many important roles in society. They might, in principle at least, be able to produce any intellectual work that could be produced by a human. They might drive cars, fly airplanes, write news, or service customers. In fact, today many of those tasks are routinely performed by programs. Robot journalists, working for Associated Press and other agencies, write many of the stories released today, autonomous cars are in the news every day, and automatic piloting of airplanes has been in use for many years. Virtual assistants that can schedule meetings on your behalf are offered on the Web by a number of companies, including Clara Labs and x.ai. A virtual assistant, such as Siri or Cortana, is included in the operating system of almost every new cellular phone and can be used by anyone, even though virtual assistants still have many shortcomings and even though there are many requests that they aren't capable of processing.

Even without a physical presence outside a computer, a digital person can write letters, buy and sell stocks, make telephone calls, manipulate bank accounts, and perform many other activities characteristic of humans that do not require physical intervention. Therefore, there is a strong economical motivation to develop digital persons and to use them to foster economic activity. In many respects, digital persons will not be at a disadvantage in relation to natural persons. They will be able to launch media campaigns, address the press, appear on TV, and influence public opinion much as a natural person might.

What should be the rights of digital persons? Should they be granted personhood and all that comes with it? Could they be owned by individuals or corporations, like slaves, or should they be masters of their own destiny? Could they own money and properties? Could they vote? Could they be terminated—erased from memory—if deemed no longer useful, perhaps because a new model or an upgrade, more intelligent and flexible, has become available?

The questions are not any easier if one thinks about the responsibilities of digital persons. Suppose a digital person does something illegal. Should it take the blame and the responsibility, and go to jail, whatever that may mean in the specific context? Or should the blame be attributed to the programmers who, perhaps many years ago, wrote the routines that enabled the digital person to acquire the knowledge and the personality that led to its current behavior? Should a digital person be sentenced to death if, by accident or by intent, it has caused the death of other persons, natural or otherwise? At the time of this writing, the US National Transportation Safety Board is analyzing the conditions under which a Model S Tesla automobile operating in its Autopilot mode failed to brake before impact with a white truck it hadn't detected, thereby causing the death of its human occupant. Other similar cases are likely to follow as the technologies become more widespread.

If a program passes a full-fledged Turing Test, convincing everyone that it reasons as a human does, should the prize be given to the programmers who created it, or to the program? The programmers would probably be glad to receive the prize, but the program might disagree with this option. After all, if a program is human enough to pass a Turing Test, is it not human enough to keep a share of the prize?

Although these questions seem outlandish, they are not idle musings about hypothetical entities that will never exist. Intelligent agents—digital minds—are likely to exist sooner or later, and, if they do, discussions about whether or not they are digital persons will be unavoidable. If true digital persons should come into existence, we can be sure that the challenges that will be posed to society will be formidably more complex than the ones that are at the center of our most contentious issues today.

Originals, Copies, and Duplicates

One major difference between the digital world and the physical world is that digital entities, which exist only in computer memories, can easily be copied, stored, and retrieved. In the physical world, people place significant importance on the distinction between originals and copies. A bank teller will willingly cash a signed check, but will not be happy if presented with two copies of the same check, or even with two different checks of which the second bears a scanned copy of the signature on the first. A restaurant will treat you very differently if you pay for dinner with a $50 bill than if you pay with a copy (even a good copy) of a $50 bill, and there is a huge difference in value between a painting by Picasso and a copy of it.

This state of affairs is consistent with the inherent difficulty of perfect duplication of objects in the physical world. Until very recently, exact duplication of physical objects was either impossible or very hard, and the way society works rests on the assumption that perfect copies of physical objects are difficult to obtain. Forgeries are possible, but are usually detectable and almost always illegal. Technologies for handling currency, signatures, fingerprints, and all other means of personal identification are based on the fact that exact copies of physical objects are hard to create. In particular, it is very difficult to make exact copies of human bodies, or of parts of human bodies, such as fingers or retinas, and thus biometric methods can be used to verify the identity of a natural person with almost absolute certainty. In fact, one of the most common means of identification—a handwritten signature—is based on the difficulty of reproducing the sequence of movements that a human hand makes when producing a physical signature.

But this situation, in which authentication and identification mechanisms are based on the difficulty of copying physical objects, is changing rapidly, for two reasons. The first reason is that duplicating physical objects exactly is not as difficult as it once was. The second and more important reason is that in the digital realm duplication is easy.

Technologies already under development will enable us, in the near future, to perform high-precision scans of physical objects and to assemble (perhaps with the use of nanotechnologies) any number of copies, using so-called universal assemblers. Nanotechnologies may one day make it possible to assemble objects, atom by atom, from specifications stored in digital form. The practical challenges such technologies will bring to our physical world are interesting; however, they are not at the center of the present discussion, if only because it is likely that similar and much more complex problems will show up in the digital world well before universal assemblers become available.

In the digital world, exact copies are easy to obtain. We are all familiar with the challenges that existing digital technologies create today. Software is easily duplicated, as are all sorts of media stored in digital format, such as books, music, videos, and photographs. All the industries based on creating, copying, and distributing media content, including the newspaper, book, movie, and music industries, are facing the need to adapt to a world in which exact copies of a digital object are easy to make, obtain, and distribute widely. Some business models will adapt and some will not, but the changes in these industries are so profound that very little of what was in place just ten years ago will survive, in its present form, for another ten years. Newspapers, booksellers, movie studios, songwriters, and singers are having to deal with profound changes in the way people produce and consume content.

In the digital world, there is no difference between an original and a copy. The very essence of digital entities makes it irrelevant whether some instance is an original or a copy. In some cases—for instance, when you take a digital picture and store it in the camera card—it is possible to identify exactly where the sequence of bits that constitutes a digital entity was first assembled. In other cases, it is more difficult or even impossible to identify where a digital entity came in existence—for instance, when you transmit a voice message by phone and it is recorded somewhere in the cloud, to be reproduced later, after having been sent through the air in many dispersed packets. But in none of these cases is the "original" sequence of bits distinguishable from copies made of it. The copy of a digital picture you store in your computer is as good as the version first recorded in the camera card, and any copies that you send to your friends are not only of the same quality, but, in a sense, as original as the one stored in the camera card. In these considerations I am ignoring the accumulation of errors in digital entities. These errors are rare and can be easily corrected.

The possibility of duplicating digital persons raises a plethora of complex philosophical questions. In fact, one may harbor genuine doubts about the eventual feasibility and existence of digital minds. However, it is completely safe to assume that, if digital minds ever come into existence, it will be practicable to back them up, store them for future retrieval, and, in general, copy them exactly an arbitrary number of times. Independently of the exact techniques used to create and develop digital minds, the underlying computational support will probably be based on standard digital storage, which by its very nature allows exact copying an arbitrary number of times.

Dualists will always be able to argue that the essence of a mind cannot be copied, even if the digital support of the system that generates the mind can. Believers in the theory that quantum phenomena play a significant role in the emergence of consciousness will take solace in the fact that the quantum configuration of a system cannot be copied without interfering with the original, whereas a digital entity can be copied without disturbing the configuration of the original. However, there is no evidence that any of these beliefs holds true. It is highly likely that digital minds, if they ever come into existence, will be trivially easy to copy and duplicate. It is hard to grasp even a partial set of the implications of this possibility, since many scenarios that are not possible in the physical world become not only possible but common in the digital world.

Consider the possibility of a direct copy of a digital person. Now imagine that a technology that makes non-destructive mind uploading possible will be developed, sometime in this century or the next, and that an uploaded mind obtains the status of a digital person.

First, it will be necessary to deal with the question of how to treat the digital person and the biological human who served as the template for the uploading. Immediately after duplication, the behavior of the two persons—the digital one and the biological one—will be identical or at least very similar. The copy, which will live in some sort of virtual environment, will want to have access to some resources that will enable it to carry on with its life. Maybe it will want money to buy computer time to run the emulation, or to pay for other resources, physical or digital. The biological human, having (presumably) agreed to be scanned and copied into the memory of a computer, may view this desire with some sympathy and may even be willing to share his or her resources with the digital version. Both are exactly the same person, with the same memories, the same feelings, and the same history, up to the moment of duplication, at which time they will begin to diverge. Divergence will occur slowly and accumulate over time. One can imagine that duplication, in this case, can be similar either to a self-divorce, with the possessions divided between the original and the copy, or to a marriage, with the possessions jointly owned by the original and the copy.

Maybe non-destructive mind uploading will be legally possible only after a suitable binding contract is established between the two future copies of the person who will be uploaded. There is, of course, the small difficulty that these two copies do not exist at the moment the upload operation is initiated; however, that isn't a serious concern, because both copies will be indistinguishable from the original. Immediately after the uploading operation, both copies will have a common past but divergent futures. This contract could then be written and registered, before the uploading, in effect binding both future versions of the person signing the contract. Such a contract, signed by the natural person before the uploading, would enforce the future distribution of assets, and would presumably be binding for the two (identical) persons after the duplication, since their past selves agreed and signed. It is difficult to anticipate whether or not such a mechanism would work, but it seems to be the most logical way to ensure that a discussion about the rights and assets of the two entities doesn't ensue. One may counter that such a contract may be deemed unacceptable by one (or both) of the future persons, but legal mechanisms may be developed to ensure the right framework for this type of procedure, that was, after all, agreed to by both parties.

It is also possible that mind uploading, should it come to pass, will have to use a destructive technology. Owing to the limitations of scanning technology, it may be necessary to destroy the original brain in order to obtain a detailed copy that then can be emulated. This case would be less complex than the preceding one; there wouldn't really be duplication, and deciding who owns the assets would presumably be simpler. Heirs might claim that their predecessor is now dead; barring that possibility, however, it is arguable that the uploaded digital person should retain (or perhaps inherit) the rights and assets of the original biological person. In writing this, I am assuming the legal rights and responsibilities of digital persons would have been established by the time the technology became feasible.

Once a digital mind exists, obtained either by uploading or by some other method, duplication of digital persons will be easy. Duplication and instantiation of a new copy of a digital person is a fairly straightforward process; it would be possible even with present-day technology if a computer-executable version of a digital person could be provided. The challenges are similar to the challenges with mind uploading, but in this case it makes no sense to assume the original is destroyed in the process. Again, a legal contract, binding both instances of the digital person, may be required before duplication takes place. The difference between duplicating a digital mind and uploading a biological mind is that in the former case there exists a complete symmetry, since neither of the two copies of the digital mind can claim to be more original than the other. In the case of mind uploading, the biological person who was the basis for the copy may have a case that he or she is the original, and should be allowed to decide about the distribution of assets and even about termination of the copy.

The questions are not much simpler if, instead of mind uploading, one considers the other possibilities for mind design. If digital persons have rights, handling duplication processes will always be tricky. The essential difficulty stems from one crucial difference: There is no branching in the line that takes one biological person from birth to death, so the definition of personhood stays with the same individual from the moment of birth until the moment of death. Even though the atoms in a human being's body are constantly being replaced by new ones, and even though most of a human being's cells are replaced every few years, a human being retains a personality that is essentially unchanged, and there is no doubt, to society at least, that he or she is the same person he or she was a few years ago. Someone who committed a crime and who later argues in self-defense that he is now a different person would not get easily away from his responsibilities. The question becomes different, however, if one considers digital minds, because in that case the flow of time is less predictable—in fact, it is possible to make time go backward or forward in a limited but important sense.

Time Travel, Backups, and Rollbacks

The possibility of time travel has been a subject of many novels, movies, and science-fiction plots. Although some people still believe time travel may one day, by some mechanism, become possible, we are stuck with the fact that, except for relativistic effects, the only way we can travel through time is at the normal rate of one second per second. This is to say that, in the world we know and experience, what is possible is to travel forward in time at exactly the same speed as everyone else and the same speed as the rest of the physical world that surrounds us.

The theory of relativity predicts that, when an object moves in relation to another object, time will slow down for both objects—an effect that will be stronger when the objects are moving at very high speeds relative to one another. This time dilation has been observed and measured many times and, in fact, leads to different rates between global positioning satellites (GPS) clocks and Earth clocks, a difference that requires correction in order to make the GPS system work. By the same token, an astronaut traveling in a spaceship at a speed close to that of light might see many centuries elapse in the outside world during his or her lifetime. Barring such a mechanism for time travel (which is still beyond the reach of science and technology), we are stuck with the fact that traveling in time is not a reality in our daily lives in the physical world. Things are a bit different when one considers the possibilities open to digital persons whose lives and minds are emulated in a computer.

The emulation of a digital mind need not proceed at the usual pace of physical time. Depending only on how much computational power is available, one may decide to run the emulation of a virtual world in real time, or at a speed faster than real time, or at a speed slower than real time. To facilitate the connection with the real world, it is reasonable to assume that emulation of digital minds will usually be run at the same speed as real time. This would enable an easier interaction with the physical world, something that probably will be valuable for a long time.

For a digital person (particularly one who has experienced the physical world), it would be strange to observe time pass, in the physical world, at a speed very different from subjective personal time. Such a difference in speeds, however, may be necessary, or even desirable, for a number of reasons. One may imagine, for instance, that lack of computational resources will require the emulation to be run at 1/100 of real time. In such a case, for each subjective day elapsed, an entire season would elapse in the physical world. In a week, the digital person would see two years go by. Conversations with natural persons in the physical world, and other interactions too, would have to be slowed down to a rate that would make it possible for both sides to communicate. Other phenomena, too, would feel very awkward for the digital person. Subjectively, the digital person would be traveling forward in time at the rate of 100 seconds for each elapsed second. This would be sufficient to observe children, in the outside world, grow visibly every day and friends getting old at an uncommonly high speed.

The opposite might also be true. If sufficient computer power were available, digital minds could be emulated at a rate faster than real time, in effect slowing down external time from the point of view of the digital person. In an emulation ran 100 times the speed of real time, the digital person would see two years of his or her life go by while in the physical world only a week elapsed. Again, conversations with external persons, and other interactions, would be awkward and difficult, requiring long pauses between sentences so that the biological persons in the outside world would be able to process what was said by the digital person, who would be talking at a much higher speed.

More radical forms of forward time travel are also possible. For instance, a digital person might decide to jump forward in time ten years instantaneously. All that would be required would be stopping the emulation for the specified period of time, which would require no more than the flip of a switch. For the digital person, no time would have elapsed at all when, ten years later, the emulation was restarted. In the blink of an eye, ten years would have gone by in the outside world, from the subjective point of view of the digital person whose emulation was stopped.

The existence of digital minds and digital worlds also make possible a somewhat limited version of time travel to the past. I have already alluded to the fact that digital minds, as any other digital entities, can be copied at will. This means that it is possible to copy the status of the digital mind to some storage medium at any given time and to recover it later—perhaps many years later. This rather simple action makes it possible to perform regular backups of a digital person and to recover, at any time in the future, the digital person exactly in the same state it was when the backup was made. In this way, it becomes possible to bring back anyone from any time in the past, if a backup from that particular time is available.

As anyone familiar with computer technology knows, it is quite feasible to perform regular and inexpensive backups of a computer by means of a technique called *incremental backups*. More sophisticated than normal backups, incremental backups store only the changes from the previous backup and can be used to recover the state of a computer in any specific point in time. If something like this technology is used to guarantee the recovery of a digital person from any point in time in the past, then it is not only possible but easy to go back in time, from the subjective point of view of the digital person.

This possibility has a number of interesting consequences. First, the actual death of a digital person will never happen unless it is specifically desired, because it is always possible to go back in time and reinstate a younger version of the person from a backup. In Charles Stross' novel *Glasshouse*, fatal duels are common occurrences, culturally and socially accepted, because the duelists always have the possibility of being restored from a backup performed before the duel. The worst thing that can happen is to lose the memories from the time between the last backup and the duel—something that is seen as inconvenient but not dramatic.

Other possibilities now thought of as in the realm of science fiction become possible. One of these possibilities is for a digital person to have a conversation with a younger or older version of itself. For instance, if a digital person desires to let a younger version of itself know a few useful facts about its future life, all it need do is reinstate the younger version from a backup and arrange for a private chat with it, perhaps over a few pints of beer. From then on, things get a bit more confusing. If the younger version is to be able to make use of that information, it will have to go on with life in the possession of this valuable piece of advice from a uniquely informed source. But then there will be two copies of the same person in existence, with different ages. Or perhaps the older person will regret so much the course taken by its life that it will choose to be terminated and let the younger version become the new person, now enlightened with the wisdom received from an older self. In practice, this represents a way to correct decisions made in the past that a digital person regrets later in life. For instance, after a failed romantic relationship, a digital person may decide to terminate itself and to reinstate a previous backup copy, starting with a clean slate. In the movie *Eternal Sunshine of the Spotless Mind*, a sophisticated procedure is able to selectively delete memories that no longer are wanted. Unlike in the movie, restoring from a previous backup is equivalent to removing all the memories from the intervening period (and not a selected few), between the moment of the backup and the instant the digital person goes back in time.

The questions and challenges raised by these strange possibilities don't end here; in fact, they are almost endless. Suppose that a digital person with a clean slate commits a serious crime and is sentenced to termination.

Should a backup of that person, obtained before the crime, be allowed to be restored? If restored, should it be liable for the crime its copy committed in its subjective future, which is, however, the past for the rest of the world? One should keep in mind that the nature of serious crimes will change significantly if digital persons come to exist. In the presence of backups, murder of digital persons would not be very serious, as long as the murdered person has a relatively recent backup. In fact, murder might be seen as a quite common event, with minor consequences—perhaps the equivalent of a slap in the face in today's world. And suicide would be ineffective, as a digital person can be easily reinstated from an earlier backup.

All these strange new possibilities, and many I have not addressed, will remain open to discussion in the decades to come, as the technologies needed to create digital persons are developed. However, it will be useful if advances in our understanding of these questions are achieved before we are faced with the actual problems—something that may happen sooner than is now expected, at least in some particular aspects.

Living in Virtual Reality

At this point, it is worthwhile discussing in some detail how exactly digital minds interact with the physical world, if they do so at all. So far, I have glossed over the practical difficulties involved in such interaction. I have described some of the technical issues related to the emulation of brains, or to the design of other computational agents that may give rise to digital minds, but I have not addressed the important question of interaction with the outside world. This is an important question for uploaded minds and for neuromorphic systems, because a precise and realistic emulation of a brain will require detailed interaction with the physical world. Without such interaction, it will not be practicable to establish the appropriate connections between neurons that create experiences, memories, and emotions.

Obviously, inputs from the outside world could be fed to the optic and auditory nerves, through cameras and microphones, in order to stimulate the senses of vision and audition. Other sensory nerves, responsible for other types of sensations, could also be stimulated to provide a digital mind with the senses of touch, smell, and taste, so as to make life in a virtual world similar to life in the physical world. The engineering challenges that would have to be surmounted to endow a digital person with a complete set of stimuli are probably as complex as the challenges that would have to be surmounted to create a digital mind, but they are essentially technological challenges. Although the basic five senses are the best-known senses, humans have other senses, which provide them with varied inputs from the physical world. Arguably, however, a detailed emulation that includes complete versions of the five senses is already a sufficiently big challenge to deserve discussion.

In the other direction, a digital mind would have to be able to influence the outside world by controlling the muscles that generate speech and movement. Those muscles are controlled by motor neurons, which receive inputs from other neurons (typically in the spinal cord) and transmit signals to the muscles to produce movement.

The technologies required to achieve an effective simulation of this complex process are the subjects of active research, since they will be also useful to help disabled people control prosthetic limbs and for many other purposes. Brain plasticity (the ability of the brain to adapt to new stimuli) will come in handy when a newly created digital mind is provided with inputs from an unfamiliar source, such as a digital camera or a microphone. This plasticity will enable the brain to adapt and to learn how to generate the right patterns of nerve firing. Brain adaptation has been extensively observed and used in experimental settings in which blind patients have had their neural cortices stimulated directly by visual inputs (Dobelle, Mladejovsky, and Girvin 1974) and deaf patients have their cochlea stimulated directly by auditory inputs (Crosby et al. 1985). There is still the challenge of using this information in the (simulated) motor neurons to interact with the outside world and to move the sensorial devices to the places where the action is.

The more obvious possibility is that a digital mind inside a computer will interact with the outside world through a robot. Advanced robotics technologies will be able to reproduce faithfully the intended movements and interactions of a digital mind, much like the artificially intelligent robots in the movie *Ex Machina*. Humanoid or not, robots will enable digital persons to interact with the physical world. An array of technologies required to enable robots with a complete set of sensors and actuators that will mimic the sensors and actuators humans have will take many decades to develop, but there is no intrinsic technical difficulty that couldn't be addressed by advances in robotics. Researchers working on brain-machine interfaces (BMIs) have already achieved many significant results, including devices that enable quadriplegic patients to control limbs using only thoughts. Such technologies will one day be used to enable normal humans to control computers, perhaps replacing existing interfaces. They can also be used by digital minds to control robots in the physical world. BMI researchers have also developed technologies that can be used to perform direct stimulation of brain areas in order to create sensations equivalent to touch and other senses. Such technologies could be used to provide digital minds with realistic sensations of immersion in the physical world.

The second possibility is less obvious but significantly more flexible. A digital person could live in a completely simulated digital world in which the inputs to the senses and the actions caused by the muscles were simulated in a virtual environment. Most readers probably are familiar with many types of virtual reality commonly used in computer games. Today's most advanced videogames already have an impressive level of realism. Videogames that simulate soccer or basketball already achieve levels of realism that are comparable to real-world images. When all the interaction occurs on a computer screen the sensation of immersion is incomplete, but soon more immersive interfaces will come closer to fooling the senses completely.

Virtual reality has been hailed by many as the future way to interact with computers. Virtual-reality glasses, for instance, have been proposed by a number of companies, but so far they have failed to become a significant way to interact with machines. That is due in part to the fact that getting a virtual-reality device to simulate visual inputs accurately is difficult. The brain's sensitivity to inconsistencies in movements and in rendered images leads to sickness and disorientation. However, as the technology of virtual reality develops, the quality of the interaction has improved. Oculus VR and some other companies now promise virtual-reality experiences that are essentially indistinguishable from real visual experiences. Google's innovative Cardboard project puts virtual reality in the hands of anyone who has a smartphone. For less than $20, one can acquire a Cardboard kit with which one can, in conjunction with a smartphone application, experience an immersive and very convincing virtual-reality environment.

So far, these immersive interfaces have addressed only vision and audition. Addressing other sensations, such as touch or movement, is much more difficult when dealing with physical bodies, which would have to be moved and/or put into immersive special environments. The actuators and sensors that would be required would be physically large and difficult to implement. This challenge, however, is much easier to address if there is no physical body present, but only a digital mind running in a computer. In that case, the physical sensations of movement, including acceleration, could be caused by simply acting on the right sensory nerves. The resulting sensation would be of complete and realistic immersion.

I believe that, with the advances in image rendering technologies and with a somewhat better understanding of the operation of motor neurons and sensory neurons, highly realistic versions of digital worlds will not only be possible, but will ultimately become the mechanism of choice to enable digital persons to interact with the outside world. Of course, the outside world here is a very particular one, since it will be entirely simulated within a computer. Cities filled with people, grass-covered fields, beaches with breaking waves, and snow-covered mountains could all be simulated to a level of realism that would make them, in practice, indistinguishable from the real world. Interaction with other digital minds would proceed straightforwardly, since those minds would inhabit this virtual world. Interactions with the actual physical world could also exist, although they would require additional devices. The state of the physical world, or of parts of it, could be fed into the virtual simulation by cameras and other sensors, and appropriate renderings of the humans and other agents in the physical world could be performed by the computer running the virtual-world simulation. Actual physical interaction with persons and objects in the outside physical world would remain the limiting factor; this limitation might be addressed by using robots or some sort of mechanism of synchronization between the two worlds.

The concept of a large number of digital minds living in a virtual world and interacting with one another may seem outlandish and far-fetched. However, it may be the most obvious and cost-effective way of creating conditions appropriate for digital minds. The future may, indeed, be something like the world depicted in Greg Egan's novel *Permutation City* (2010a), in which the bodies and brains of 10 million people are emulated on a single computer chip.

It is a trivial fact, but still worthwhile noting, that in a virtual world the laws of physics as we know them need not necessarily apply. In a purely virtual environment, people can travel instantaneously from one place to another, anywhere in the universe, as long as the destination can be simulated. In such a virtual world, digital persons can fly over mountains, swim to the bottom of the ocean, or walk on the surface of the Moon with no special equipment. In fact, the possibilities are endless, and I cannot anticipate even a shadow of the virtual realities that will one day be created. Some of these strange possibilities are already familiar to players of virtual-reality games.

In such a virtual world, even going back in time becomes genuinely possible. From inside a computer-generated virtual reality, it must be possible to re-run the simulation from any particular point in time. Of course, some parameter in the simulation would have to be changed; otherwise the simulation would run again in exactly the same way. Perhaps some digital persons would like to run (virtual) reality again, from a specific point in the past, with some added knowledge they didn't have the first time they went through that point. Since the whole simulated world would have to be rerun, this would correspond to a duplication of (virtual) reality, with a corresponding expense of computational resources and with the implied duplication of all digital entities in the simulated world. That would raise another set of questions that are simply too alien to discuss here.

I cannot anticipate whether digital persons would, in general, prefer to live in a realistic depiction of the outside physical world, or would instead opt for a simulation environment more flexible and free from the laws of physics that have kept humanity limited in so many aspects for millions of years. If such a virtual environment were to be created, we would truly be in the realm of science fiction. I don't believe we are equipped, at the moment, to think about the issues that would have to be addressed.

Personality Surgery and Other Improvements

So far, we have assumed that digital minds, created by uploading biological minds, would be created by means of a procedure that would ensure that their behavior would be essentially indistinguishable from that of the original minds. Synthetic minds are different, since they would have been designed in accordance with entirely different principles and since they probably would work in a way very different the way the human mind works. Neuromorphic intelligent systems would be somewhere in between. The most obvious way to guarantee that a digital copy of a biological mind behaves like the original is to make the copy as faithful as is possible. We know the technology that would be needed to create such a copy doesn't yet exist and, indeed, may never exist. However, even if the technology required to copy a mind to a digital support comes into existence, it may not be sufficient to understand and modify the emulated mind.

In the case of uploaded minds, one may imagine that the copying is done by some automatic process that simply identifies brain structures and mechanisms and copies them to the memory of a computer, which will then be able to emulate the processes taking place inside the brain. Understanding how a brain works is a different challenge, perhaps much more complex. The activity patterns in the brain that lead to a particular feeling or memory are probably complex and hard to understand from first principles. Emotions and other brain functions probably are emergent properties of complex firing patterns in the brain that cannot be easily understood, much less changed or improved.

Our understanding of how the brain works may eventually reach a point at which minds can be modified, finely tuned, improved, and changed at the owner's discretion. Manipulating the brain's structures and patterns to erase memories selectively or to improve intelligence may be possible. However, such manipulation will require a level of understanding of brain behavior that doesn't necessarily result automatically from an ability to create digital minds.

There has been significant progress in brain science in recent decades, and such progress is likely to continue at an ever-accelerating pace. But even if it is possible to gain detailed knowledge of the mechanisms that control the inner workings of the brain, we must keep in mind that every human brain is different from every other. Understanding the general mechanisms is not the same thing as understanding the behavior of a specific brain. General chemical imbalances and other large-scale processes that lead to illnesses could be addressed much as they are addressed by today's medication-based techniques. To achieve that, it would suffice to change some parameters in the emulation corresponding to the environment variables that one wants to change—variables that, in real life, are controlled by drugs, by surgery, or by other macro-scale processes. However, detailed neuron-level surgery capable of wiping out a single memory or a single behavior, if such surgery were to become possible, would require very detailed knowledge of the behavior of every individual brain. This may be much more difficult than obtaining a general understanding of the general mechanisms that underlie brain behavior.

There is one factor that will make it much easier to achieve a detailed understanding of the behavior of specific minds. The behavior of a digital mind can be observed with any required level of detail, whereas with a biological mind instrumentation and physical limitations set hard limits on what can be observed. Even with the advances in instrumentation I have discussed in chapter 9, it is still not possible to obtain detailed information about the firing of individual neurons in a working brain. A digital mind, obtained by emulating a human brain in a computer, would make available for inspection every single firing pattern, every single connection, and every single phenomenon occurring in a working brain. Such information could be used by researchers not only to understand how existing minds work but also, presumably, to learn how to design more powerful minds. This raises the interesting possibility that digital minds might be able to improve themselves, since they would, in principle at least, be able to inspect and understand their own behavior with an unprecedented level of detail.

12 Speculations

The possibilities addressed in the chapters above, far-fetched as they may seem, raise the possibility that future technological developments may be even more unpredictable than one might expect from recent history. In this final chapter, I discuss a few possible long-term directions for the evolution of technology. These are educated guesses or, if you prefer, wild speculations.

The Technological Singularity

Bootstrapping is a term that may have originated a couple of centuries ago to speak of an imagined ability to lift oneself up by one's own bootstraps. It implies the ability of a process to sustain itself, and to keep reinforcing itself, once it has been started.

Digital minds have a number of advantages over biological minds. They can be copied an arbitrary number of times, they can be run at different speeds (depending on how much computational power is available), and they can be instrumented and changed in ways that are not applicable to biological minds. It is reasonable to assume that digital minds, once in place, could be used to advance our understanding of intelligence. These simple advantages of digital minds alone would make them important in the development of advanced AI technologies.

The basic idea that an intelligent system, running in a computer, can bootstrap itself in order to become more and more intelligent, without the need for human intervention, is an old one and has appeared in a number of forms in scientific works and in science fiction. This idea is sometimes called *seed AI,* a term used to imply that what is needed to create an ever-accelerating process is to create the first artificially intelligent system sufficiently smart to understand and improve itself. Although no such system exists, or is even planned, it is reasonable to assume that digital minds, if they ever exist, will play an important role in the further development of the technologies that led to their very own existence.

In 1965, Irving John Good (a mathematician who worked with Alan Turing at Bletchley Park, contributing to the British effort to break the Germans' Enigma codes) coined the term *intelligence explosion*:

> Let an ultraintelligent machine be defined as a machine that can far surpass all the intellectual activities of any man however clever. Since the design of machines is one of these intellectual activities, an ultraintelligent machine could design even better machines; there would then unquestionably be an 'intelligence explosion,' and the intelligence of man would be left far behind. Thus the first ultraintelligent machine is the last invention that man need ever make, provided that the machine is docile enough to tell us how to keep it under control. (Good 1965)

Good believed that if machines could even slightly surpass human intelligence, they would surely be able to improve their own designs in ways not foreseen by their human designers, recursively augmenting their intellects past the level of human intelligence.

The possibility of bootstrapping, by which intelligent systems would lead to even more intelligent systems in an accelerating and self-reinforcing cycle, is one of the factors that have led many people to believe we will eventually observe a discontinuity in the way society is organized. Such a discontinuity has been called *the technological singularity* or simply *the singularity*. The term refers to a situation in which technological evolution leads to even more rapid technological evolution in a rapidly accelerating cycle that ends in an abrupt and radical change of the whole society. Intelligent machines are viewed as playing a major role in this cycle, since they could be used to design successively more intelligent machines, leading to levels of intelligence and technological capacity that would rapidly exceed human levels. Since this would lead to a situation in which humans would no longer be the dominant intelligence on Earth, the technological singularity would be an event beyond which the evolution of technology and society would become entirely unpredictable.

In mathematics, a singularity is a point at which a function or some other mathematical object reaches some exceptional or non-defined value or fails to have some specific smoothness property. For instance, the function $1/x$ has a singularity when $x = 0$, because at that point the function has no defined value. As x approaches 0 while taking positive values, the value of the function moves rapidly toward infinity.

This idea of a mathematical singularity inspired the use of the term *technological singularity* to denote a point at which there will be a lack of continuity in the technological evolution of mankind. The first person to use the word *singularity* in this context may have been the mathematician and physicist John von Neumann. In his 1958 tribute to von Neumann, Stanislaw Ulam describes a conversation in which he quotes von Neumann as having said that "the ever-accelerating progress of technology and changes in the mode of human life, which gives the appearance of approaching some essential singularity in the history of the race beyond which human affairs, as we know them, could not continue" (Ulam 1958).

The expression *technological singularity* was popularized by Vernor Vinge, a scientist and a science-fiction author who has argued that a number of other causes could also cause or contribute to the singularity (Vinge 1993), and by Ray Kurzweil, who wrote a number of books that address what he sees as the coming singularity (1990, 2000, 2005).

It is important to note that technologies other than digital minds, natural or synthetic, could also lead to a technological singularity. Radical improvements in biology and medicine could lead to an ability to replace or regenerate human bodies and human brains, creating the possibility of eternal or almost eternal life. Such a situation would certainly disrupt many social conventions, but would not, in my mind, really correspond to a singularity, since society would probably be able to adapt, more or less smoothly, to a situation in which death was no longer inevitable. Significant advances in nanotechnologies, which make it possible to create from raw atoms any physical object, would also change many of the basic tenets of society, perhaps leading to a society in which any object could be obtained by anyone. However, legislation and other restrictions would probably ensure enough continuity so that human life, as we know it, would still be understandable to us.

To my mind, none of these technologies really compares, in its disruptive potential, to a technology capable of creating digital minds. Digital persons, living in a digital world, bound only by the limitations imposed by available computer power, would create an entirely different society—a society so different from the one we currently have that would justify the application of the term *singularity*. Increasing computational power as much as is possible would lead to new computing technologies and architectures, perhaps directing a large fraction of the Earth's resources to create ever-more-powerful computers. Science fiction and futurologists have envisaged a situation in which the whole Earth would be turned into a computer, and have even created a term for a substance that would be used to create computers as efficiently as would be possible from the resources available. This substance, *computronium*, has been defined by some as a substance that extracts as much computational power as possible from the matter and energy available. Such a material would be used to create the ultimate computer, which would satisfy the ever-increasing needs of a civilization running in a virtual world. In fact, the evolution of computers, leading to ever smaller and more powerful computing devices, seems to take us in the direction of computronium, even though at present only a small fraction of the Earth's mass and energy is dedicated to making and operating computers. However, as transistors become smaller and smaller, and more and more resources are dedicated to computations, aren't we progressing steadily toward creating computronium?

There is no general agreement as to whether the singularity might come to pass, much less on when. Optimists imagine it arriving sometime in the first half of the twenty-first century, in time to rescue some people alive today from the inevitable death brought by aging. Other, less optimistic estimates put it several hundred years in the future. In 2008, *IEEE Spectrum*, the flagship publication of the Institute of Electrical and Electronics Engineers, dedicated a special issue to the topic of the singularity. Many scientists, engineers, visionaries, and science-fiction writers discussed the topic, but no clear conclusions were drawn.

Many arguments have been leveled against the idea of the singularity. We have already met the most obvious and perhaps the strongest arguments. They are based on the idea that minds cannot be the result of a computational process and that consciousness and true intelligence can only result from the operation of a human brain. According to these arguments, any other intelligence, resulting from the operation of a computer, will always be a minor intelligence, incapable of creating the self-accelerating, self-reinforcing, process required to create the singularity.

Other arguments are based on the fact that most predictions of future technologies—underwater cities, nuclear fusion, interplanetary travel, flying automobiles, and so on—have not materialized. Therefore, the prediction of a technological singularity is no more likely to come true than many of these other predictions, simply because technology will take other routes and directions.

Many opponents of the idea of a singularity base their arguments on sociological and economic considerations. As more and more jobs disappear and are replaced by intelligent agents and robots, they argue, there will be less and less incentive to create more intelligent machines. Severe unemployment and reduced consumer demand will destroy the incentive to create the technologies required to bring about the singularity. Jared Diamond argues in his 2005 book *Collapse: How Societies Choose to Fail or Succeed* that cultures eventually collapse when they exceed the sustainable carrying capacity of their environment. If that is true, then our civilization may not evolve further in the direction of the singularity.

A final line of argument is based on the idea that the rate of technological innovation has already ceased to rise and is actually now declining (Modis 2006). Evidence for this decline in our capacity to innovate is the fact that computer clock speeds are not increasing at the same rate and the mounting evidence that physical limitations will ultimately put a stop to Moore's Law. If this argument is correct, and such a decline in innovation capacity is indeed happening, then the second half of the twentieth century and the first decades of the twenty-first will have witnessed the highest rate of technological change ever to occur.

I don't have a final, personal opinion as to whether or not the singularity will happen. Although I believe that the evolution of technology follows a tendency that is essentially exponential, I recognize that such an analysis is true only when considered in broad strokes. Detailed analysis of actual technological innovations would certainly identify large deviations from such an exponential tendency. It is also true that exponential functions have no singularities. They grow faster and faster as time goes by, but they never reach a mathematical singularity.

However, it is also true, in general, that exponential evolutions proceed so rapidly that they eventually reach some physical limit. The exponential growth of living cells, caused by their ability to reproduce, can last only as long as the physical resources are sufficient; the growth slows when it reaches some excessively high level. Maybe the "exponential" evolution of technology we have been observing in recent decades will reach a limit, and the rate of innovation will then decrease, for technological or social reasons, leading to a more stable and predictable evolution of society.

However, even if the technological singularity doesn't happen, I believe that changes will continue to accumulate at such a high rate that society will, by the end of the present century, have changed so drastically that no one alive today would recognize it. I have already mentioned Arthur C. Clarke's third law, which states that any sufficiently advanced technology is indistinguishable from magic. Airplanes, cars, computers, and cellular phones would have appeared as magic to anyone living only a few hundred years ago. What wondrous technologies will the future bring? Digital minds may be only one of them!

Through the Singularity, Dead or Alive

True believers in the singularity may wish to take measures to ensure that their personalities, if not their bodies, are preserved long enough to be able to take advantage of the possibilities the singularity may bring. The most obvious and conspicuous approach is used by people who believe in cryonics, the cryopreservation of human bodies. Cryopreservation is a well-developed process whereby cells and tissues are preserved by subjecting them to cooling at very low temperatures. If the temperatures are low enough (near 80 kelvin, approximately the temperature of liquid nitrogen), enzymatic or chemical activity that may cause damage to biological tissues is effectively stopped, which makes it possible to preserve specimens for extended periods of time. The idea of cryonics is to keep human brains or whole human bodies frozen while waiting for the technologies of the future.

The challenge is to reach these low temperatures while avoiding a number of phenomena that cause extensive damage to cells. The most damaging such phenomena are cell dehydration and extracellular and intracellular ice formation. These deleterious effects can be reduced though the use of chemicals known as *cryoprotectants,* but current technology still inflicts significant damage on organs beyond a certain size.

Once the preserved material has been frozen, it is relatively safe from suffering further damage and can be stored for hundreds of years, although other effects, such as radiation-induced cell damage, may have to be taken into consideration.

With existing technology, cryopreservation of people or of whole brains is deemed not reversible, since the damages inflicted on the tissues are extensive. Attempts to recover large mammals frozen for thousands of years by simply warming them were abandoned many decades ago. However, believers in cryonics take it for granted that future technology will be able to revert the damages inflicted by the freezing process, making it possible that cryopreserved people (or brains) may someday be brought back to life.

A number of institutions and companies, including the Cryonics Institute and the Alcor Life Extension Foundation, give their members the option of having their heads or bodies frozen in liquid nitrogen for prices that start at less than $100,000—a cost that, by any standard, has to be considered very reasonable for a fair chance at resurrection. According to the statistics published, these institutions have, as of today, a few thousand members, and already have preserved a few hundred bodies in liquid nitrogen.

There is a strong ongoing debate about the feasibility of cryonics (Merkle 1992). An open letter supporting the feasibility of the technology was signed by a number of well-known scientists, but most informed people are highly skeptical and view cryonics as a dishonest scheme to extract money from uninformed (if optimistic) customers.

Other believers in the singularity, who are more optimistic, hope it will arrive in time to save people still alive today. Of these people, the most outspoken is probably Ray Kurzweil. He believes that advances in medicine and biology will make it possible for people alive today to experience the singularity during their lifetimes (Kurzweil and Grossman 2004).

In view of the complexity of the techniques required to perform either mind uploading or mind repairing on frozen brains and the enormous technological developments still required, my personal feeling is that the hopes of the two groups of people mentioned above are probably misplaced. Even if cryonics doesn't damage brain tissues too deeply, future technologies probably will not be able to use the live brain information that, in my view, will be indispensable if mind uploading ever becomes possible. If one assumes that the singularity will not arrive before the end of the twenty-first century, the hope that it will arrive on time to rescue from certain death people still alive today rests implicitly on the assumption that advances on medicine will increase life expectancy by about one year every year, something that is definitely too optimistic.

The Dangers of Super-Intelligences

As we have seen, the jury is still out on whether or not the singularity will happen sometime in the not too distant future. However, even if the singularity doesn't happen, super-human intelligences may still come into existence, by design or accident. Irving John Good's idea of an intelligence explosion may lead to systems that are much more intelligent than humans, even if the process doesn't create a technological singularity.

AI researchers, starting with Alan Turing, have many times looked forward to a time when human-level artificial intelligence becomes possible. There is, however, no particular reason to believe that there is anything special about the specific level of intelligence displayed by humans. While it is true that humans are much more intelligent that all other animals, even those with larger brains, it is not reasonable to expect that human intelligence sits at the maximum attainable point on the intelligence scale.

Super-human intelligences could be obtained either by greatly speeding up human-like reasoning (imagine a whole-brain emulator running 100.000 times the speed of real time), by pooling together large numbers of coordinated human-level intelligences (natural or artificial), by supplying a human-level intelligence with enormous amounts of data and memory, or by developing some yet-unknown new forms of intelligence. Such super-human intelligences would be, to human intelligence, as human intelligence is to chimpanzee-level intelligence. The survival of chimpanzees, as of the other animals, now depends less on them than on us, the dominant form of life on Earth. This is not because we are stronger or more numerous, but only because we are more intelligent and have vastly superior technologies.

If a super-human intelligence ever develops, will we not be as dependent on it as chimpanzees now are on us? The very survival of the human race may one day depend on how kindly such a super-intelligence looks upon humanity. Since such a super-human intelligence will have been created by us, directly or indirectly, we may have a chance of setting things up so that such an intelligence serves only the best interests of humanity. However, this is easier said than done.

In 1942, Isaac Asimov proposed the three laws of robotics that, if implemented in all robots, would guarantee the safeguarding of human lives:

that a robot may not injure a human being or, through inaction, allow a human being to come to harm; that a robot must obey orders given to it by human beings unless such orders would conflict with the first law; and that a robot must protect its own existence as long as such protection doesn't conflict with the first law or the second. However, these laws now seem somewhat naive, as they are based on the assumption (which was common in the early years of AI research) that a set of well-defined symbolic rules would eventually lead to strong artificial intelligence. With the evolution of technology, we now understand that an artificially intelligent system will not be programmed, in minute detail, by a set of rules that fully define its behavior. Artificially intelligent systems will derive their own rules of behavior from complex and opaque learning algorithms, statistical analyses, and complex objective functions.

A super-human intelligence might easily get out of control, even if aiming for goals defined by humans. For purposes of illustration, suppose that a super-intelligent system is asked to address and solve the problem of global warming. Things may not turn out as its designers expected if the super-intelligent system determines that the most effective solution is to take human civilization back to a pre-technological state or even to eradicate humans from the surface of the Earth. You may believe this to be a far-fetched possibility, but the truth is that a truly super-intelligent system might have goals, values, and approaches very different from those held by humans. A truly super-intelligent system may be so much more intelligent than humans, and so effective at developing technologies and finding solutions, that humanity will become enslaved to its aims and means. Furthermore, its motivations might be completely alien to us—particularly if it is a synthetic intelligence (see chapter 10) with a behavior very different from the behavior of human intelligence.

The problem is made more complex by the fact the explosion of intelligence may happen in a relatively short period of time. Artificial intelligence has been under development for many decades, and mind-uploading technologies will probably take more than fifty years to develop. This would lead us to believe that an intelligence explosion, were it to happen, would take place over many decades. However, that may turn out not to be the case. A seed AI, as defined above, may be able to improve itself at a rate incommensurably faster than the rate of development of AI technologies by humans. By using large amounts of computational resources and speeding up the computation in a number of ways, a human-level artificially intelligent system may be able to increase its intelligence greatly over a period of days, hours, or even seconds. It all depends on factors that we cannot know in advance.

Some authors have argued that, before we develop the technologies that may lead to super-human intelligence, we should make sure that we will be able to control them and to direct them toward goals that benefit mankind. In his book *Superintelligence,* Nick Bostrom addresses these matters—including the potential risks and rewards—in depth, and proposes a number of approaches that would ultimately enable us to have a better chance at controlling and directing such a super-human intelligence. To me, however, it remains unclear exactly what can effectively be done to stop a truly super-human intelligence from setting its own goals and actions and, in the process, going against the interests of humanity.

Will digital minds become the servants or the masters of mankind? Only the future will tell.

Where Is Everybody?

The universe is composed of more than 100 billion galaxies. There are probably more than 200 billion stars in the Milky Way, our galaxy. If the Milky Way is a typical galaxy, the total number of stars in the universe is at least 10^{22}. Even by the lowest estimates, there are more than a billion trillion stars in the universe—a number which is, in the words of Carl Sagan, vastly larger than the number of grains of sand on Earth. We know now that many of these stars have planets around them, and many of those planets may have the conditions required to support life as we know it. We humans have wondered, for many years, whether we are alone in the universe. It is hard to conceive that such a vast place is now inhabited by only a single intelligent species, *Homo sapiens*.

Using the Drake Equation (proposed in 1961 by the radio astronomer Frank Drake to stimulate scientific discussion about the search for extraterrestrial intelligence), it is relatively easy to compute an estimate of the number of intelligent civilizations in our galaxy with long-range communication ability:

$$N = R^* \times f_p \times n_e \times f_l \times f_i \times f_c \times L,$$

where N is the number of civilizations in the Milky Way whose electromagnetic emissions are detectable, R^* is the rate of formation of stars suitable for the development of intelligent life, f_p is the fraction of those stars with planetary systems, n_e is the number of planets per solar system with an environment suitable for life, f_l is the fraction of suitable planets on which life actually appears, f_i is the fraction of life bearing planets on which intelligent life emerges, f_c is the fraction of civilizations that develop a technology that releases detectable signs of their existence into space, and L is the length of time such civilizations release detectable signals into space. Some of the factors, such as the rate of star formation and the fraction of stars with planets, are relatively easy to estimate from sky surveys and the study of exoplanets. The other factors are more difficult or even impossible to estimate accurately.

Three of the factors in the Drake Equation are particularly hard to estimate with any accuracy. One is the fraction of planets amenable to life that actually develop life. There is no reliable way to estimate this number, although many researchers believe it is highly probable that life will develop on a planet if the right conditions exist. This belief is supported in part by the fact that life appeared on Earth shortly (in geological terms) after the right conditions were present. This is seen as evidence that many planets that can support life will eventually see it appear. However, the anthropic principle gets somewhat in the way of this argument. After all, our analysis of the history of appearance of life on Earth is very biased, since only planets that have developed life have any chance of supporting the intelligent beings making the analysis.

The second hard-to-estimate factor is the fraction of life-bearing planets that actually develop intelligent life. On the basis of the idea that life will eventually develop intelligence as an effective tool for survival, some estimates of this factor propose a value close to 1. In the other direction, the argument states that the value must be very low, because there have been probably more than a billion species on Earth and only one of them developed intelligence. This argument is not exactly true, because there were multiple species in the genus *Homo,* all of them presumably rather intelligent, although all but one are now extinct (Harari 2014). Still, considerable uncertainty remains about the right value for this factor, which can be anywhere in the range between 0 and 1. As Caleb Scharf described clearly in his captivating book *The Copernicus Complex,* inhabiting the only known planet that evolved intelligent life makes it particularly difficult to obtain accurate and unbiased estimates for these two factors.

The third difficult-to-estimate factor is the length of time a civilization lasts and emits communication signals into space. We don't have even one example that could enable us to estimate the duration of a technological civilization, nor do we have physical principles to guide us. There is no way to know whether our technological space-faring civilization will last 100 years or 100 million years. Historically, specific civilizations have lasted between a few decades and a few hundred years, but it is hard to argue that history is a good guide in this respect, given the technological changes that took place in the twentieth century. In fact, I don't believe we are equipped with the right tools to reason about the future of technological civilizations lasting for millions of years. There is simply no reasonable way to extrapolate, to that length of time, our experience of a technological civilization that has been in place for only a few hundred years.

If one assumes a reasonable probability that life develops whenever the right conditions are met on a planet and a conservative but not overly pessimistic value for f_i (the fraction of life-bearing planets that develop intelligent life), such as 1 percent (Scharf 2014), the crucial factor determining the number of living technological civilizations in the galaxy is, in fact, L, the length of time a technological civilization endures. If one assumes a value of only a few hundred years, it is highly likely that there are only a few technological civilizations, and perhaps only one, in the galaxy. If, on the other hand, a technological civilization lasts for millions of years, there may be hundreds of thousands of such civilizations in the galaxy.

This leads us to one difficult question regarding extra-terrestrial civilizations, known as Fermi's Paradox: Where are they? That question was posed in 1950 by Enrico Fermi when he questioned why we have never seen evidence of advanced extraterrestrial civilizations, if many of them exist in our galaxy.

Answers to this question have to be based on the Rare Earth Hypothesis, on the argument that intelligent life is a very uncommon occurrence on life-bearing planets, or on the argument that communicating civilizations last only for short spans of time. Each of these three arguments supports a very low value for at least one of the difficult-to-estimate factors in the Drake Equation.

The Rare Earth Hypothesis (Ward and Brownlee 2000) states that the origin of life and the evolution of biological complexity require a highly improbable combination of events and circumstances and are very likely to have occurred only once, or only a small number of times, in the galaxy. This hypothesis leads to very small values of the factors n_e and f_l, (the number of planets per star that support life and the fraction of those that will actually develop life, respectively). If the Rare Earth Hypothesis is true, then there are only a few planets in the galaxy, maybe only one, that have developed life. There are many reasons why planets similar to Earth may be very rare. These reasons include long-term instability of planetary orbits and solar systems, fairly frequent planetary cataclysms, and the low likelihood that all the things necessary to support life are present on a particular planet.

I have already presented some of the discussions about f_i, the probability that a life-bearing planet will develop intelligent life. Although the answer to Fermi's Paradox may lie in the fact that there are many life-bearing planets but few with intelligent life, most see this possibility as unlikely, since it would imply that the universe is teaming with life but intelligent life evolved only on Earth.

The third explanation is based on the idea that civilizations tend not to last long enough to communicate with one another, since they remain viable for only a few hundred years. This explanation may have some bearing on the central topic of this book. There are a number of reasons why a technological civilization that communicates with the outside world and develops space travel may last only a few hundred years.

One possible explanation is that such a civilization will destroy itself, either because is exhausts the resources of the planet, because it develops tensions that cannot be handled, or because it develops a hostile super-intelligence. Such an explanation holds only if one believes that the collapse of a civilization leads to a situation in which the species that created the civilization becomes extinct or permanently pre-technological. This doesn't seem very likely, in view of our knowledge of human history, unless such an event is caused by an unprecedented situation, such as the creation of a hostile super-intelligence not interested in space communication.

The alternative explanation is that such a civilization doesn't become extinct, but evolves to a more advanced state in which it stops communicating with the outside world and doesn't develop space travel.

Paradoxically, digital minds can provide an explanation to Fermi's Paradox. It may happen that all sufficiently advanced civilizations end up developing mechanisms for the creation of virtual realities so rich that they develop their own internal, synthetic universes and no longer consider space travel necessary or even desirable. Virtual realities, created in digital computers or other advanced computational supports by sufficiently advanced civilizations, may become so rich and powerful that even the most fascinating of all explorations, interstellar travel, comes to be thought irrelevant and uninteresting.

If such an explanation holds true, it is possible that many technological civilizations do exist in the galaxy, living in virtual realities, hives of collective minds running in unthinkably powerful computers, creating their own universes and physical laws. In fact, it is even possible that we live inside one such virtual reality, which we call, simply, the universe.

Further Reading

Chapter 1

The topics covered in this chapter cover the recent and not-so-recent history of technology. For some interesting views on this history, see *Sapiens: A Brief History of Humankind* by Yuval Noah Harari (Harper, 2014) and *Guns, Germs, and Steel* by Jared Diamond (Chatto & Windus, 1997).

Chapter 2

Johan Goudsblom's *Fire and Civilization* (Viking, 1992) and Richard Wrangham's *Catching Fire: How Cooking Made Us Human* (Basic Books, 2009) provide interesting accounts of the prehistoric use of fire. Kevin Kelly's *What Technology Wants* (Viking Books, 2010) provides a view in several respects aligned with my own views on the evolution and the future of technology. William Rosen's *The Most Powerful Idea in the World: A Story of Steam, Industry, and Invention* (Random House, 2010) provides an excellent account of the radical changes brought in by the industrial revolution. James Gleick's *The Information: A History, a Theory, a Flood* (Fourth Estate, 2011) accounts for the many reasons information became the dominant factor of modern life. Jorge Luís Borges' short story "The Library of Babel" can be found in English translation in *Ficciones* (Grove Press, 1962).

Chapter 3

Of the many books covering the story of science in the twentieth century, I recommend Robert Oerter's *The Theory of Almost Everything* (Plume, 2006), Bill Bryson's *A Short History of Nearly Everything* (Broadway Books, 2003),

Stephen Hawking's *A Brief History of Time* (Bantam Books), and Amir Aezel's *Entanglement* (Plume, 2003). Bruce Hunt's *The Maxwellians* (Cornell University Press, 1991) presents in detail the way Maxwell's ideas were developed by his followers. Michael Riordan and Lillian Hoddeson's *Crystal Fire: The Invention of the Transistor and the Birth of the Information Age* (Norton, 1997) tells, in detail, how the transistor changed the world. Richard Feynman's *QED: The Strange Theory of Light and Matter* (Princeton University Press, 1985) may still be the most accessible entry point to quantum physics. John Barrow's *The Constants of Nature* (Random House, 2010) gives a fascinating account of the many numbers that encode the secrets of the universe. Robert Lucky's *Silicon Dreams: Information, Man, and Machine* (St. Martin's Press, 1989) is an informed discussion of the technologies that created the information age. Herbert Taub and Donald Schilling's *Digital Integrated Electronics* (McGraw-Hill, 1977) is still a useful reference for those who want to understand the way digital circuits are designed. John Hennessy and David Patterson's *Computer Architecture: A Quantitative Approach* (Morgan-Kaufmann, 1990) remains the authoritative source on computer architecture.

Chapter 4

The Universal Machine (Springer, 2012), by Ian Watson, is an excellent introduction to the history of computing and the perspectives of the evolution of computers. Another perspective can be obtained from George Dyson's *Turing's Cathedral: The Origins of the Digital Universe* (Pearson, 2012). Michael Sipser's *Introduction to the Theory of Computation* (PWS, 1997) introduces the reader gently to the subtleties of languages, Turing machines, and computation. *Introduction to Algorithms* (MIT Press, 2001; third edition 2009), by Thomas Cormen, Charles Leiserson, Ronald Rivest, and Clifford Stein, remains the reference on algorithms and data structures. *Computers and Intractability: A Guide to the Theory of NP-Completeness* (Freeman, 1979), by Michael Garey and David Johnson, is an indispensable companion for anyone working on complexity. *Algorithms on Strings, Trees, and Sequences* (Cambridge University Press, 1997), by Dan Gusfield, is a comprehensive graduate-level treatise on algorithms for the manipulation of strings.

Chapter 5

Artificial Intelligence: A Modern Approach (Prentice-Hall, 1995; third edition 2010), by Stuart Russell and Peter Norvig, remains the reference text in artificial intelligence, despite the enormous number of alternatives. Elaine Rich's *Artificial Intelligence* (McGraw-Hill, 1983) provides a good idea of the attempts to approach the problem of artificial intelligence with essentially symbolic approaches, although it is now dated. *Perceptrons* (MIT Press, 1969), by Marvin Minsky and Seymour Papert, is still the final word on what a single perceptron can and cannot do, and is of historical interest. The two volumes of *Parallel Distributed Processing*, by David Rumelhart, James McClelland, and the PDP Research Group (MIT Press, 1986) were at the origin of the modern connectionist approach to artificial intelligence, and remain relevant. *The Master Algorithm* (Allen Lane, 2015), by Pedro Domingos, is an accessible and enthralling introduction to the topic of machine learning. Other, more technical approaches can be found in *The Elements of Statistical Learning* (Springer, 2009), by Trevor Hastie, Robert Tibshirani, and Jerome Friedman, in *Machine Learning* (McGraw-Hill, 1997), by Tom Mitchell, in *Probabilistic Reasoning in Intelligent Systems: Networks of Plausible Inference* (Morgan Kaufmann, 1988), by Judea Pearl, and in *Reinforcement learning: An introduction* (MIT Press, 1998), by Richard Sutton and Andrew Barto.

Chapter 6

The Selfish Gene (Oxford University Press, 1976), *The Blind Watchmaker* (Norton, 1986), and *The Ancestor's Tale* (Mariner Books, 2004), by Richard Dawkins, are only some of the books that influenced my views on evolution. Another is *The Panda's Thumb* (Norton, 1980), by Stephen Jay Gould. *Genome: The Autobiography of a Species in 23 Chapters* (Fourth Estate, 1999) and *Nature via Nurture* (Fourth Estate, 2003), both by Matt Ridley, offer stimulating accounts of the way genes control our lives. *The Cell: A Molecular Approach* (ASM Press, 2000), by Geoffrey Cooper, is a reference on cellular organization. Daniel Dennett's book *Darwin's Dangerous Idea* (Simon & Schuster, 1995) is probably one of the main influences on the present book.

Chapter 7

Of the many books that tell the story of the race for the human genome, James Watson's *DNA: The Secret of Life* (Knopf, 2003) and J. Craig Venter's *A Life Decoded: My Genome—My Life* (Penguin, 2007) are probably the ones with which to get started. J. Craig Venter's *Life at the Speed of Light* (Little, Brown, 2013) continues the story into the dawn of synthetic biology. Another, more equidistant view of this race can be found in *The Sequence: Inside the Race for the Human Genome* (Weidenfeld & Nicolson, 2001), by Kevin Davies. An account of the impact of bioinformatics in today's society can be found in Glyn Moody's *The Digital Code of Life* (Wiley, 2004). Technical introductions to bioinformatics include *Bioinformatics: The Machine Learning Approach* (MIT Press, 1998), by Pierre Baldi and Søren Brunak, and *An Introduction to Bioinformatics Algorithms* (MIT Press, 2004), by Neil Jones and Pavel Pevzner.

Chapter 8

This chapter was influenced most by David Hubel's *Eye, Brain, and Vision* (Scientific American Library, 1988) and David Marr's *Vision* (Freeman, 1982). Bertil Hille's *Ionic Channels of Excitable Membranes* (Sinauer, 1984) remains a reference on the topic of membrane behavior in neurons. J. A. Scott Kelso's *Dynamic Patterns: The Self-Organization of Brain and Behavior* (MIT Press, 1995) is one of the few books that attempt to find general organizing principles for the brain. Steven Pinker's *How the Mind Works* (Norton, 1997) and Stanislas Dehaene's *Consciousness and the Brain: Deciphering How the Brain Codes Our Thoughts* (Viking Penguin, 2014) are interesting attempts at explaining how the brain works, despite all the limitations imposed by our ignorance.

Chapter 9

Michio Kaku's *The Future of the Mind* (Penguin, 2014) includes accessible and entertaining description of the many directions being pursued in brain research. Steven Rose's *The Making of Memory: From Molecules to Mind* (Bantam, 1992) and Sebastian Seung's *Connectome: How the Brain's Wiring Makes Us Who We Are* (Houghton Mifflin Harcourt, 2012) are respectively a classic and a recent account of the efforts researchers have been pursuing to understand how the brain works. Olaf Sporns' *Networks of the Brain* (MIT Press, 2011) provides a detailed account of the methods used to study brain networks.

Chapter 10

Gödel, Escher, Bach: An Eternal Golden Braid (Harvester Press, 1979), by Douglas Hofstadter, *The Emperor's New Mind* (Oxford University Press, 1989), by Roger Penrose, *The Mind's I: Fantasies and Reflections on Self and Soul* (Bantam Books, 1981), by Douglas Hofstadter and Daniel Dennett, and *Brainchildren* (MIT Press, 1998), by Dennett, are unavoidable references for those who care about brain and mind issues. Of the many books about consciousness and free will, I recommend Susan Blackmore's *Consciousness: A Very Short Introduction* (Oxford University Press, 2005), Tor Nørretranders' *The User Illusion: Cutting Consciousness Down to Size* (Penguin, 1991), Daniel Dennett's *Freedom Evolves* (Penguin, 2003) and *Consciousness Explained* (Little, Brown, 1991), Robert Ornstein's *The Evolution of Consciousness: The Origins of the Way We Think* (Touchstone, 1991), and Sam Harris' *Free Will* (Simon & Schuster, 2012). Eric Drexler's classic *Engines of Creation: The Coming Era of Nanotechnology* (Anchor Books, 1986) is still a good reference on the topic of nanotechnology. Feng-Hsiung Hsu's *Behind Deep Blue: Building the Computer That Defeated the World Chess Champion* (Princeton University Press, 2002) provides an excellent account of the effort behind the creation of a champion chess computer. Finally, Greg Egan's *Zendegi* (Gollancz, 2010) looks so realistic that it blurs the line between science fiction and reality.

Chapter 11

Many of the moral questions discussed in these chapters are covered by Daniel Dennett's previously mentioned books and by his *Brainstorms* (MIT Press, 1981). Economic, social, philosophical, and technological issues are covered extensively in Nick Bostrom's though-provoking and fact-filled *Superintelligence* (Oxford University Press, 2014). Other topics of these chapters are also addressed by science-fiction books, among which I recommend Charles Stross' *Accelerando* (Penguin, 2005) and *Glasshouse* (Penguin, 2006),

Greg Egan's *Permutation City: Ten Million People on a Chip* (Harper, 1994), and Neal Stephenson's *The Diamond Age: Or, A Young Lady's Illustrated Primer* (Bantam Books, 1995) and *Snow Crash* (Bantam Books, 1992).

Chapter 12

On the evolution of civilizations, the future of mankind, and the possibility of the singularity, I was influenced by Jared Diamond's *Collapse: How Societies Choose to Fail or Succeed* (Penguin, 2005) and, of course, by Ray Kurzweil's *The Age of Spiritual Machines* (Orion, 1999) and *The Singularity Is Near* (Penguin, 2005). Peter Ward and Donald Brownlee's *Rare Earth: Why Complex Life Is Uncommon in the Universe* (Copernicus Books, 2000) presents the arguments for the rarity of life in the universe. Caleb Scharf's *The Copernicus Complex* (Penguin, 2014) is probably the most comprehensive reference on the quest for extraterrestrial life.

References

Agrawal, Manindra, Neeraj Kayal, and Nitin Saxena. 2004. PRIMES Is in P. *Annals of Mathematics* 160 (2): 781–793.

Almeida, Luis B. 1989. "Backpropagation in Perceptrons with Feedback." In *Neural Computers*, ed. R. Eckmiller and C. von der Malsburg. Springer.

Altschul, Stephen F., Warren Gish, Webb Miller, Eugene W. Myers, and David J. Lipman. 1990. "Basic Local Alignment Search Tool." *Journal of Molecular Biology* 215 (3): 403–410.

Anger, Andreas M., Jean-Paul Armache, Otto Berninghausen, Michael Habeck, Marion Subklewe, Daniel N. Wilson, and Roland Beckmann. 2013. "Structures of the Human and Drosophila 80S Ribosome." *Nature* 497 (7447): 80–85.

Avery, Oswald T., Colin M. MacLeod, and Maclyn McCarty. 1944. "Studies on the Chemical Nature of the Substance Inducing Transformation of Pneumococcal Types Induction of Transformation by a Desoxyribonucleic Acid Fraction Isolated from Pneumococcus Type III." *Journal of Experimental Medicine* 79 (2): 137–158.

Azevedo, Frederico A. C., Ludmila R. B. Carvalho, Lea T. Grinberg, José Marcelo Farfel, Renata E. L. Ferretti, Renata E. P. Leite, Roberto Lent, and Suzana Herculano-Houzel. 2009. "Equal Numbers of Neuronal and Nonneuronal Cells Make the Human Brain an Isometrically Scaled Up Primate Brain." *Journal of Comparative Neurology* 513 (5): 532–541.

Baeza-Yates, Ricardo, and Berthier Ribeiro-Neto. 1999. *Modern Information Retrieval*. ACM Press.

Barrell, B. G., A. T. Bankier, and J. Drouin. 1979. "A Different Genetic Code in Human Mitochondria." *Nature* 282 (5735): 189–194.

Beggs, John, and Dietmar Plenz. 2003. "Neuronal Avalanches in Neocortical Circuits." *Journal of Neuroscience* 23 (35): 11167–11177.

Belliveau, J. W., D. N. Kennedy, R. C. McKinstry, B. R. Buchbinder, R. M. Weisskoff, M. S. Cohen, J. M. Vevea, T. J. Brady, and B. R. Rosen. 1991. "Functional Mapping of the Human Visual Cortex by Magnetic Resonance Imaging." *Science* 254 (5032): 716–719.

Berger, Hans. 1929. Über das Elektrenkephalogramm des Menschen. *European Archives of Psychiatry and Clinical Neuroscience* 87 (1): 527–570.

Berman, Helen M, John Westbrook, Zukang Feng, Gary Gilliland, T. N. Bhat, Helge Weissig, Ilya N. Shindyalov, and Philip E. Bourne. 2000. The Protein Data Bank. *Nucleic Acids Research* 28 (1): 235–242.

Bianconi, Eva, Allison Piovesan, Federica Facchin, Alina Beraudi, Raffaella Casadei, Flavia Frabetti, Lorenza Vitale, Maria Chiara Pelleri, Simone Tassani, and Francesco Piva. 2013. "An Estimation of the Number of Cells in the Human Body." *Annals of Human Biology* 40 (6): 463–471.

Bissell, Chris. 2007. Historical Perspectives: The Moniac A Hydromechanical Analog Computer of the 1950s. *IEEE Control Systems Magazine* 27 (1): 69–74.

Blakemore, Colin, and Grahame F. Cooper. 1970. "Development of the Brain Depends on the Visual Environment." *Nature* 228 (5270): 477–478.

Bliss, T. V., and G. L. Collingridge. 1993. "A Synaptic Model of Memory: Long-Term Potentiation in the Hippocampus." *Nature* 361 (6407): 31–39.

Bliss, T. V., and Terje Lømo. 1973. "Long-Lasting Potentiation of Synaptic Transmission in the Dentate Area of the Anaesthetized Rabbit Following Stimulation of the Perforant Path." *Journal of Physiology* 232 (2): 331–356.

Blumer, A., A. Ehrenfeucht, D. Haussler, and M. Warmuth. 1987. "Occam's Razor." *Information Processing Letters* 24 (6): 377–380.

Blumer, A., and R. L. Rivest. 1988. Training a 3-Node Neural Net Is NP-Complete. *Advances in Neural Information Processing Systems* suppl. 1:494–501.

Boas, David A., Anders M. Dale, and Maria Angela Franceschini. 2004. "Diffuse Optical Imaging of Brain Activation: Approaches to Optimizing Image Sensitivity, Resolution, and Accuracy." *Neuroimage* 23 (suppl. 1): S275–S288.

Bock, Davi D., Wei-Chung Allen Lee, Aaron M. Kerlin, Mark L. Andermann, Greg Hood, Arthur W. Wetzel, Sergey Yurgenson, et al. 2011. "Network Anatomy and *In Vivo* Physiology of Visual Cortical Neurons." *Nature* 471 (7337): 177–182.

Borůvka, O. 1926. "O Jistém Problému Minimálnıım." *Praca Moravske Prirodovedecke Spolecnosti*, no. 3: 37–58.

Bostrom, Nick. 2014. *Superintelligence: Paths, Dangers, Strategies*. Oxford University Press.

Boyden, Edward S., Feng Zhang, Ernst Bamberg, Georg Nagel, and Karl Deisseroth. 2005. "Millisecond-Timescale, Genetically Targeted Optical Control of Neural Activity." *Nature Neuroscience* 8 (9): 1263–1268.

Breiman, Leo, Jerome Friedman, Charles J. Stone, and Richard A. Olshen. 1984. *Classification and Regression Trees*. CRC Press.

Brin, S., and L. Page. 1998. "The Anatomy of a Large-Scale Hypertextual Web Search Engine." In *Proceedings of the Seventh International World Wide Web Conference*. Elsevier.

Brocks, Jochen J., Graham A. Logan, Roger Buick, and Roger E. Summons. 1999. "Archean Molecular Fossils and the Early Rise of Eukaryotes." *Science* 285 (5430): 1033–1036.

Brodmann, K. 1909. *Vergleichende Lokalisationslehre der Groshirnrinde*. Barth.

Brownell, Gordon L., and William H. Sweet. 1953. "Localization of Brain Tumors with Positron Emitters." *Nucleonics* 11 (11): 40–45.

Bugalho, Miguel M. F., and Arlindo L. Oliveira. 2008. "An Evaluation of the Impact of Side Chain Positioning on the Accuracy of Discrete Models of Protein Structures." *Advances in Bioinformatics and Computational Biology* 5167: 23–34.

Caton, Richard. 1875. The Electric Currents of the Brain. *British Medical Journal* 2:278–278.

Charlesworth, Brian, and Deborah Charlesworth. 2009. "Darwin and Genetics." *Genetics* 183 (3): 757–766.

Church, Alonzo. 1936. "An Unsolvable Problem of Elementary Number Theory." *American Journal of Mathematics* 58 (2): 345–363.

Clancy, B., R. B. Darlington, and B. L. Finlay. 2001. "Translating Developmental Time across Mammalian Species." *Neuroscience* 105 (1): 7–17.

Clarke, Arthur C. 1968. *A Space Odyssey*. Hutchinson.

Cohen, David. 1968. "Magnetoencephalography: Evidence of Magnetic Fields Produced by Alpha-Rhythm Currents." *Science* 161 (3843): 784–786.

Compeau, Phillip E. C., Pavel A. Pevzner, and Glenn Tesler. 2011. "How to Apply de Bruijn Graphs to Genome Assembly." *Nature Biotechnology* 29 (11): 987–991.

Cook, S. A. 1971. "The Complexity of Theorem-Proving Procedures." In *Proceedings of the Third Annual ACM Symposium on Theory of Computing*. ACM.

Cooke, S. F., and T. V. Bliss. 2006. Plasticity in the Human Central Nervous System. *Brain* 129 (7): 1659–1673.

Cooper, Geoffrey M., and Robert E. Hausman. 2000. *The Cell*. Sinauer.

Craik, Kenneth James Williams. 1967. *The Nature of Explanation*. Cambridge University Press.

Crick, Francis, Leslie Barnett, Sydney Brenner, and Richard J. Watts-Tobin. 1961. "General Nature of the Genetic Code for Proteins." *Nature* 192 (4809): 1227–1232.

Crombie, Alistair Cameron. 1959. *Medieval and Early Modern Science.*, volume II. Harvard University Press.

Crosby, Peter A., Christopher N. Daly, David K. Money, James F. Patrick, Peter M. Seligman, and Janusz A. Kuzma. 1985. Cochlear Implant System for an Auditory Prosthesis. US Patent 4532930.

Crosson, Bruce, Anastasia Ford, Keith M. McGregor, Marcus Meinzer, Sergey Cheshkov, Xiufeng Li, Delaina Walker-Batson, and Richard W. Briggs. 2010. "Functional Imaging and Related Techniques: An Introduction for Rehabilitation Researchers." *Journal of Rehabilitation Research and Development* 47 (2): vii–xxxiv.

Darwin, Charles. 1859. *On the Origin of Species by Means of Natural Selection*. Murray.

Davidson, Donald. 1987. Knowing One's Own Mind. *Proceedings and Addresses of the American Philosophical Association* 60 (3): 441–458.

Davis, Martin, Yuri Matiyasevich, and Julia Robinson. 1976. Hilbert's Tenth Problem: Diophantine Equations: Positive Aspects of a Negative Solution. *Proceedings of Symposia in Pure Mathematics* 28:323–378.

Davis, Martin, Hilary Putnam, and Julia Robinson. 1961. The Decision Problem for Exponential Diophantine Equations. *Annals of Mathematics* 74 (3): 425–436.

Dawkins, Richard. 1976. *The Selfish Gene*. Oxford University Press.

Dawkins, Richard. 1986. *The Blind Watchmaker: Why the Evidence of Evolution Reveals a Universe without Design*. Norton.

Dawkins, Richard. 2010. *The Ancestor's Tale: A Pilgrimage to the Dawn of Life*. Orion.

Denk, Winfried, and Heinz Horstmann. 2004. Serial Block-Face Scanning Electron Microscopy to Reconstruct Three-Dimensional Tissue Nanostructure. *PLoS Biology* 2 (11): e329.

Dennett, Daniel C. 1991. *Consciousness Explained*. Little, Brown.

Dennett, Daniel C. 1995. *Darwin's Dangerous Idea*. Simon and Schuster.

Dennett, Daniel C. 2003. *Freedom Evolves*. Penguin.

Descartes, René. 1641. *Meditationes de Prima Philosophia*.

Diamond, Jared. 1997. *Guns, Germs, and Steel*. Norton.

Diamond, Jared. 2005. *Collapse: How Societies Choose to Fail or Succeed*. Penguin.

Dias, A., M. Bianciardi, S. Nunes, R. Abreu, J. Rodrigues, L. M. Silveira, L. L. Wald, and P. Figueiredo. 2015. "A New Hierarchical Brain Parcellation Method Based on Discrete Morse Theory for Functional MRI Data." In *IEEE 12th International Symposium on Biomedical Imaging.*

Dijkstra, Edsger W. 1959. "A Note on Two Problems in Connexion with Graphs." *Numerische Mathematik* 1 (1): 269–271.

Dobelle, William H., M. G.Mladejovsky, and J. P. Girvin. 1974. "Artificial Vision for the Blind: Electrical Stimulation of Visual Cortex Offers Hope for a Functional Prosthesis." *Science* 183 (4123): 440–444.

Domingos, Pedro. 1999. "The Role of Occam's Razor in Knowledge Discovery." *Data Mining and Knowledge Discovery* 3 (4): 409–425.

Domingos, Pedro. 2015. *The Master Algorithm: How the Quest for the Ultimate Learning Machine Will Remake Our World.* Basic Books.

Domingos, Pedro, and Michael Pazzani. 1997. "On the Optimality of the Simple Bayesian Classifier under Zero-One Loss." *Machine Learning* 29 (2–3): 103–130.

Drexler, K. Eric. 1986. *Engines of Creation: The Coming Era of Nanotechnology.* Anchor.

Drubach, Daniel. 2000. *The Brain Explained.* Prentice-Hall.

Edmondson, James C., and Mary E. Hatten. 1987. "Glial-Guided Granule Neuron Migration in Vitro: A High-Resolution Time-Lapse Video Microscopic Study." *Journal of Neuroscience* 7 (6): 1928–1934.

Egan, Greg. 2010a. *Permutation City.* Hachette.

Egan, Greg. 2010b. *Zendegi.* Hachette.

Einstein, Albert. 1905a. Über die von der Molekularkinetischen Theorie der Wärme geforderte Bewegung von in ruhenden Flüssigkeiten suspendierten Teilchen. *Annalen der Physik* 17 (8): 549–560.

Einstein, Albert. 1905b. Über einen die Erzeugung und Verwandlung des Lichtes betreffenden heuristischen Gesichtspunkt. *Annalen der Physik* 17 (6): 132–148.

Einstein, Albert. 1905c. Zur Elektrodynamik bewegter Körper. *Annalen der Physik* 17 (10): 891–921.

Einstein, Albert, Boris Podolsky, and Nathan Rosen. 1935. "Can Quantum-Mechanical Description of Physical Reality Be Considered Complete?" *Physical Review* 47 (10): 777–780.

Feynman, Richard Phillips. 1985. *QED: The Strange Theory of Light and Matter.* Princeton University Press.

Fiers, Walter, Roland Contreras, Fred Duerinck, Guy Haegeman, Dirk Iserentant, Jozef Merregaert, W. Min Jou, et al. 1976. "Complete Nucleotide Sequence of Bacteriophage MS2 RNA: Primary and Secondary Structure of the Replicase Gene." *Nature* 260 (5551): 500–507.

Fischl, Bruce, David H. Salat, Evelina Busa, Marilyn Albert, Megan Dieterich, Christian Haselgrove, Andre van der Kouwe, et al. 2002. "Whole Brain Segmentation: Automated Labeling of Neuroanatomical Structures in the Human Brain." *Neuron* 33 (3): 341–355.

Fisher, Ronald A. 1936. "Has Mendel's Work Been Rediscovered?" *Annals of Science* 1 (2): 115–137.

Fleischner, Herbert. 1990. *Eulerian Graphs and Related Topics*. Elsevier.

Floyd, Robert W. 1962. Algorithm 97: Shortest Path. *Communications of the ACM* 5 (6): 345.

Franklin, Allan, A. W. F. Edwards, Daniel J. Fairbanks, and Daniel L. Hartl. 2008. *Ending the Mendel-Fisher Controversy*. University of Pittsburgh Press.

Friedman, Jerome, Trevor Hastie, and Robert Tibshirani. 2001. *The Elements of Statistical Learning*. Springer.

Fuller, Richard Buckminster, and John McHale. 1967. *World Design Science Decade, 1965–1975*. World Resources Inventory.

Garey, Michael R., and David S. Johnson. 1979. *Computers and Intractability: A Guide to NP-Completeness*. Freeman.

Gaur, Albertine. 1992. *A History of Writing*. University of Chicago Press.

Glover, Paul, and Sir Peter Mansfield. 2002. "Limits to Magnetic Resonance Microscopy." *Reports on Progress in Physics* 65 (10): 1489–1511.

Gödel, Kurt. 1931. Über formal unentscheidbare Sätze der Principia Mathematica und verwandter Systeme I. *Monatshefte für Mathematik und Physik* 38 (1): 173–198.

Good, Irving John. 1965. "Speculations Concerning the First Ultraintelligent Machine." *Advances in Computers* 6 (99): 31–83.

Goudsblom, Johan. 1992. *Fire and Civilization*. Viking.

Gould, Stephen Jay. 1996. *Full House: The Spread of Excellence from Plato to Darwin*. Harmony Books.

Gratton, Gabriele, and Monica Fabiani. 2001. "Shedding Light on Brain Function: The Event-Related Optical Signal." *Trends in Cognitive Sciences* 5 (8): 357–363.

Grosberg, Richard K., and Richard R. Strathmann. 2007. "The Evolution of Multicellularity: A Minor Major Transition?" *Annual Review of Ecology, Evolution, and Systematics* 38 (1): 621–654.

Gusfield, Dan. 1997. *Algorithms on Strings, Trees, and Sequences: Computer Science and Computational Biology*. Cambridge University Press.

Hagmann, Patric, Leila Cammoun, Xavier Gigandet, Reto Meuli, Christopher J. Honey, Van J. Wedeen, and Olaf Sporns. 2008. Mapping the Structural Core of Human Cerebral Cortex. *PLoS Biology* 6 (7): e159.

Hagmann, Patric, Lisa Jonasson, Philippe Maeder, Jean-Philippe Thiran, Van J. Wedeen, and Reto Meuli. 2006. "Understanding Diffusion MR Imaging Techniques: From Scalar Diffusion-Weighted Imaging to Diffusion Tensor Imaging and Beyond 1." *Radiographics* 26 (suppl. 1): S205–S223.

Hanson, Robin. 2008. Economics of the Singularity. *IEEE Spectrum* 45 (6): 45–50.

Harari, Yuval Noah. 2014. *Sapiens: A Brief History of Humankind*. Harper.

Harmand, Sonia, Jason E. Lewis, Craig S. Feibel, Christopher J. Lepre, Sandrine Prat, Arnaud Lenoble, Xavier Boës, et al. 2015. "3.3-Million-Year-Old Stone Tools from Lomekwi 3, West Turkana, Kenya." *Nature* 521 (7552): 310–315.

Hawrylycz, Michael J., Ed S. Lein, Angela L. Guillozet-Bongaarts, Elaine H. Shen, Lydia Ng, Jeremy A. Miller, and Louie N. van de Lagemaat, et al. 2012. "An Anatomically Comprehensive Atlas of the Adult Human Brain Transcriptome." *Nature* 489 (7416): 391–399.

Hebb, Donald Olding. 1949. *The Organization of Behavior: A Neuropsychological Theory*. Wiley.

Heckemann, Rolf A., Joseph V. Hajnal, Paul Aljabar, Daniel Rueckert, and Alexander Hammers. 2006. "Automatic Anatomical Brain MRI Segmentation Combining Label Propagation and Decision Fusion." *NeuroImage* 33 (1): 115–126.

Helmstaedter, Moritz, Kevin L. Briggman, Srinivas C. Turaga, Viren Jain, H. Sebastian Seung, and Winfried Denk. 2013. "Connectomic Reconstruction of the Inner Plexiform Layer in the Mouse Retina." *Nature* 500 (7461): 168–174.

Herraez, Angel. 2006. "Biomolecules in the Computer: Jmol to the Rescue." *Biochemistry and Molecular Biology Education* 34 (4): 255–261.

Higham, Tom, Katerina Douka, Rachel Wood, Christopher Bronk Ramsey, Fiona Brock, Laura Basell, Marta Camps, Alvaro Arrizabalaga, Javier Baena, and Cecillio Barroso-Ruíz. 2014. "The Timing and Spatiotemporal Patterning of Neanderthal Disappearance." *Nature* 512 (7514): 306–309.

Hilbert, David. 1902. "Mathematical Problems." *Bulletin of the American Mathematical Society* 8 (10): 437–479.

Hirsch, Helmut V. B., and D. N. Spinelli. 1970. "Visual Experience Modifies Distribution of Horizontally and Vertically Oriented Receptive Fields in Cats." *Science* 168 (3933): 869–871.

Hobbes, Thomas. 1651. *Leviathan: Or, The Matter, Form, and Power of a Commonwealth Ecclesiastical and Civil*. Routledge.

Hodgkin, Alan L, and Andrew F. Huxley. 1952. "A Quantitative Description of Membrane Current and Its Application to Conduction and Excitation in Nerve." *Journal of Physiology* 117 (4): 500–544.

Hofstadter, Douglas R., and Daniel C. Dennett. 2006. *The Mind's I: Fantasies and Reflections on Self and Soul*. Basic Books.

Hornik, Kurt, Maxwell Stinchcombe, and Halbert White. 1989. "Multilayer Feedforward Networks Are Universal Approximators." *Neural Networks* 2 (5): 359–366.

Hounsfield, Godfrey N. 1973. "Computerized Transverse Axial Scanning (Tomography): Part 1. Description of System." *British Journal of Radiology* 46 (552): 1016–1022.

Hsieh, Jiang. 2009. *Computed Tomography: Principles, Design, Artifacts, and Recent Advances*. Wiley.

Hubel, David H. 1988. *Eye, Brain, and Vision*. Scientific American Library.

Hubel, David H., and Torsten N. Wiesel. 1962. "Receptive Fields, Binocular Interaction and Functional Architecture in the Cat's Visual Cortex." *Journal of Physiology* 160 (1): 106–154.

Hubel, David H., and Torsten N. Wiesel. 1968. "Receptive Fields and Functional Architecture of Monkey Striate Cortex." *Journal of Physiology* 195 (1): 215–243.

Hume, David. 1748. *An Enquiry Concerning Human Understanding*. Millar.

Hunt, Bruce J. 1991. *The Maxwellians*. Cornell University Press.

Jaynes, Julian. 1976. *The Origin of Consciousness in the Breakdown of the Bicameral Mind*. Houghton Mifflin.

Kasthuri, Narayanan, Kenneth Jeffrey Hayworth, Daniel Raimund Berger, Richard Lee Schalek, José Angel Conchello, Seymour Knowles-Barley, Dongil Lee, et al. 2015. "Saturated Reconstruction of a Volume of Neocortex." *Cell* 162 (3): 648–661.

Kelly, Kevin. 2010. *What Technology Wants*. Penguin.

Kelso, J. A. Scott. 1997. *Dynamic Patterns: The Self-Organization of Brain and Behavior*. MIT Press.

Knight, Tom. 2003. *Idempotent Vector Design for Standard Assembly of Biobricks*. MIT Artificial Intelligence Laboratory.

Knoll, Andrew H., Emmanuelle J. Javaux, D. Hewitt, and P. Cohen. 2006. "Eukaryotic Organisms in Proterozoic Oceans." *Philosophical Transactions of the Royal Society B: Biological Sciences* 361 (1470): 102310–38.

Kononenko, Igor. 1993. Inductive and Bayesian Learning in Medical Diagnosis. *Applied Artificial Intelligence* 7 (4): 317–337.

Kunkel, S., M. Schmidt, J. Eppler, H. Plesser, G. Masumoto, J. Igarashi, S. Ishii, et al. 2014. Spiking Network Simulation Code for Petascale Computers. *Frontiers in Neuroinformatics* 8 (10): 1–23.

Kurzweil, Ray. 1990. *The Age of Intelligent Machines*. MIT Press.

Kurzweil, Ray. 2000. *The Age of Spiritual Machines: When Computers Exceed Human Intelligence*. Penguin.

Kurzweil, Ray. 2005. *The Singularity Is Near: When Humans Transcend Biology*. Penguin.

Kurzweil, Ray, and Terry Grossman. 2004. *Fantastic Voyage: Live Long Enough to Live Forever*. Rodale.

Laird, John E., Allen Newell, and Paul S. Rosenbloom. 1987. "SOAR: An Architecture for General Intelligence." *Artificial Intelligence* 33 (1): 1–64.

Lander, Eric S., Lauren M. Linton, Bruce Birren, Chad Nusbaum, Michael C. Zody, Jennifer Baldwin, Keri Devon, et al. 2001. "Initial Sequencing and Analysis of the Human Genome." *Nature* 409 (6822): 860–921.

Laureyn, W., C. Liu, C. K. O. Sullivan, V. C. Ozalp, O. Nilsson, C. Fermér, K. S. Drese, et al. 2007. "Lab-on-Chip for the Isolation and Characterization of Circulating Tumor Cells." In *Proceedings of the 29th Annual Conference of the IEEE on Engineering in Medicine and Biology Society*. IEEE.

Lauterbur, Paul C. 1973. "Image Formation by Induced Local Interactions: Examples Employing Nuclear Magnetic Resonance." *Nature* 242 (5394): 190–191.

Le Bihan, Denis, and E Breton. 1985. "Imagerie de Diffusion in-Vivo Par Résonance Magnétique Nucléaire." *Comptes-Rendus de l'Académie des Sciences* 93 (5): 27–34.

Leitão, A. C., A. P. Francisco, R. Abreu, S. Nunes, J. Rodrigues, P. Figueiredo, L. L. Wald, M. Bianciardi, and L. M. Silveira. 2015. "Techniques for Brain Functional Connectivity Analysis from High Resolution Imaging." In *Proceedings of the 6th Workshop on Complex Networks*. Springer.

Libet, Benjamin, Curtis A. Gleason, Elwood W. Wright, and Dennis K. Pearl. 1983. Time of Conscious Intention to Act in Relation to Onset of Cerebral Activity (Readiness-Potential): The Unconscious Initiation of a Freely Voluntary Act. *Brain* 106 (3): 623–642.

Lilienfeld, Julius. 1930. Method and Apparatus for Controlling Electric Currents. US Patent 1745175.

Lømo, T. 1966. "Frequency Potentiation of Excitatory Synaptic Activity in Dentate Area of Hippocampal Formation." *Acta Physiologica Scandinavica* 68 (suppl. 277): 128.

Lucas, John R. 1961. Minds, Machines and Gödel. *Philosophy (London, England)* 36 (137): 112–127.

Magee, John F. 1964. "How to Use Decision Trees in Capital Investment." *Harvard Business Review* 42 (5): 79–96.

Marasco, Addolorata, Alessandro Limongiello, and Michele Migliore. 2013. "Using Strahler's Analysis to Reduce up to 200-Fold the Run Time of Realistic Neuron Models." *Scientific Reports* 3 (2934): 1–7.

Marchant, Jo. 2008. *Decoding the Heavens: A 2,000-Year-Old-Computer—and the Century-Long Search to Discover Its Secrets*. Heinemann.

Maxwell, James Clerk. 1865. A Dynamical Theory of the Electromagnetic Field. *Philosophical Transactions of the Royal Society of London* 155:459–512.

Maxwell, James Clerk. 1873. *A Treatise on Electricity and Magnetism*. Clarendon.

Mazziotta, J., A. Toga, A. Evans, P. Fox, J. Lancaster, K. Zilles, R. Woods, et al. 2001. A Probabilistic Atlas and Reference System for the Human Brain: International Consortium for Brain Mapping (ICBM). *Philosophical Transactions of the Royal Society of London* 356 (1412): 1293–1322.

McCulloch, Warren S., and Walter Pitts. 1943. "A Logical Calculus of the Ideas Immanent in Nervous Activity." *Bulletin of Mathematical Biophysics* 5 (4): 115–133.

McVean, G. A., and 1000 Genomes Project Consortium. 2012. "An Integrated Map of Genetic Variation from 1,092 Human Genomes." *Nature* 491 (7422): 56–65.

Menabrea, Luigi Federico, and Ada Lovelace. 1843. "Sketch of the Analytical Engine Invented by Charles Babbage." *Scientific Memoirs* 3: 666–731.

Mendel, Gregor. 1866. "Versuche über Pflanzenhybriden." *Verhandlungen des Naturforschenden Vereines in Brunn* 4: 3–47.

Merkle, Ralph C. 1992. "The Technical Feasibility of Cryonics." *Medical Hypotheses* 39 (1): 6–16.

Metzker, Michael L. 2010. "Sequencing Technologies—the next Generation." *Nature Reviews Genetics* 11 (1): 31–46.

Michalski, Ryszard S., Jaime G. Carbonell, and Tom M. Mitchell. 2013. *Machine Learning: An Artificial Intelligence Approach*. Springer.

Miedaner, Terrel. 1978. *The Soul of Anna Klane: A Novel*. Ballantine Books.

Minsky, Marvin. 1961. "Microscopy Apparatus." US Patent 3013467.

Minsky, Marvin, and Seymour Papert. 1969. *Perceptrons*. MIT Press.

Modis, Theodore. 2006. "The Singularity Myth." *Technological Forecasting and Social Change* 73 (2): 104–112.

Mondragon, A., S. Subbiah, S. C. Almo, M. Drottar, and S. C. Harrison. 1989. "Structure of the Amino-Terminal Domain of Phage 434 Represser at 2.0 Å Resolution." *Journal of Molecular Biology* 205 (1): 189–200.

Moore, Gordon E. 1965. "The Density of Transistors Assembled on a Micro Chip Doubles Every 12 Months." *Electronics* 38 (8): 114–117.

Myers, Eugene W., Granger G. Sutton, Art L. Delcher, Ian M. Dew, Dan P. Fasulo, Michael J. Flanigan, Saul A Kravitz, et al. 2000. "A Whole-Genome Assembly of Drosophila." *Science* 287 (5461): 2196–2204.

Nagel, Georg, Martin Brauner, Jana F. Liewald, Nona Adeishvili, Ernst Bamberg, and Alexander Gottschalk. 2005. "Light Activation of Channelrhodopsin-2 in Excitable Cells of *Caenorhabditis elegans* Triggers Rapid Behavioral Responses." *Current Biology* 15 (24): 2279–2284.

Nagel, Georg, Tanjef Szellas, Wolfram Huhn, Suneel Kateriya, Nona Adeishvili, Peter Berthold, Doris Ollig, et al. 2003. Channelrhodopsin-2, a Directly Light-Gated Cation-Selective Membrane Channel. *Proceedings of the National Academy of Sciences of the United States of America* 100 (24): 13940–13945.

Nagel, Thomas. 1974. "What Is It Like to Be a Bat?" *Philosophical Review* 83 (4): 435–450.

NCBI (National Center for Biotechnology Information). 2007. The Cells of Eukaryotes and Prokaryotes. https://commons.wikimedia.org/wiki/File:Celltypes.svg.

Newell, Allen, and Herbert A. Simon. 1956. "The Logic Theory Machine—A Complex Information Processing System." *IRE Transactions on Information Theory* 2 (3): 61–79.

Nirenberg, Marshall, and Philip Leder. 1964. "RNA Codewords and Protein Synthesis: The Effect of Trinucleotides upon the Binding of sRNA to Ribosomes." *Science* 145 (3639): 1399–1407.

Nørretranders, Tor. 1991. *The User Illusion*. Penguin.

Ogawa, Seiji, Tso-Ming Lee, Alan R. Kay, and David W. Tank. 1990. Brain Magnetic Resonance Imaging with Contrast Dependent on Blood Oxygenation. *Proceedings of the National Academy of Sciences of the United States of America* 87 (24): 9868–9872.

Okada, Y. 1983. "Neurogenesis of Evoked Magnetic Fields." In *Biomagnetism: An Interdisciplinary Approach*, ed. S. J. Williamson, G.-L. Romani, L. Kaufman, and I. Modena. Springer.

Olazaran, Mikel. 1996. "A Sociological Study of the Official History of the Perceptrons Controversy." *Social Studies of Science* 26 (3): 611–659.

Oliveira, Arlindo L. 1994. Inductive Learning by Selection of Minimal Complexity Representations. PhD dissertation, University of California, Berkeley.

Papert, Seymour. 1966. The Summer Vision Project. http://dspace.mit.edu/handle/1721.1/6125.

Parfit, Derek. 1984. *Reasons and Persons*. Oxford University Press.

Pauling, Linus. Robert B. Corey, and Herman R. Branson. 1951. "The Structure of Proteins: Two Hydrogen-Bonded Helical Configurations of the Polypeptide Chain." *Proceedings of the National Academy of Sciences* 37 (4): 205–211.

Penrose, Roger. 1989. *The Emperor's New Mind: Concerning Computers, Minds, and the Laws of Physics*. Oxford University Press.

Pevzner, Pavel A., Haixu Tang, and Michael S. Waterman. 2001. "An Eulerian Path Approach to DNA Fragment Assembly." *Proceedings of the National Academy of Sciences* 98 (17): 9748–9753.

Phillips, A. William. 1950. Mechanical Models in Economic Dynamics. *Economics* 17 (67): 283–305.

Pires, Ana M., and João A. Branco. 2010. "A Statistical Model to Explain the Mendel–Fisher Controversy." *Statistical Science* 25 (4): 545–565.

Post, Emil L. 1946. "A Variant of a Recursively Unsolvable Problem." *Bulletin of the American Mathematical Society* 52 (4): 264–268.

Prim, Robert Clay. 1957. "Shortest Connection Networks and Some Generalizations." *Bell System Technical Journal* 36 (6): 1389–1401.

Quinlan, J R. 1986. "Induction of Decision Trees." *Machine Learning* 1 (1): 81–106.

Ragan, Timothy, Lolahon R. Kadiri, Kannan Umadevi Venkataraju, Karsten Bahlmann, Jason Sutin, Julian Taranda, Ignacio Arganda-Carreras, et al. 2012. Serial Two-Photon Tomography for Automated *Ex Vivo* Mouse Brain Imaging. *Nature Methods* 9 (3): 255–258.

Raichle, Marcus E. 2009. A Paradigm Shift in Functional Brain Imaging. *Journal of Neuroscience* 29 (41): 12729–12734.

Raichle, Marcus E., and Mark A. Mintun. 2006. "Brain Work and Brain Imaging." *Annual Review of Neuroscience* 29: 449–476.

Ramón y Cajal, Santiago. 1904. *Textura Del Sistema Nervioso Del Hombre Y de Los Vertebrados*. Nicolas Moya.

Ramón y Cajal, Santiago. 1901. *Recuerdos de mi vida*. Fortanet.

Ranson, Stephen Walter. 1920. *The Anatomy of the Nervous System*. Saunders.

Reimann, Michael W., Costas A. Anastassiou, Rodrigo Perin, Sean L. Hill, Henry Markram, and Christof Koch. 2013. "A Biophysically Detailed Model of Neocortical Local Field Potentials Predicts the Critical Role of Active Membrane Currents." *Neuron* 79 (2): 375–390.

Rich, Elaine. 1983. *Artificial Intelligence*. McGraw-Hill.

Riedlinger, Reid, Ron Arnold, Larry Biro, Bill Bowhill, Jason Crop, Kevin Duda, Eric S. Fetzer, et al. 2012. "A 32 Nm, 3.1 Billion Transistor, 12 Wide Issue Itanium® Processor for Mission-Critical Servers." *IEEE Journal of Solid-State Circuits* 47 (1): 177–193.

Rivest, R. L., A. Shamir, and L. Adleman. 1978. "A Method for Obtaining Digital Signatures and Public-Key Cryptosystems." *Communications of the ACM* 21 (2): 120–126.

Rosenblatt, Frank. 1958. "The Perceptron: A Probabilistic Model for Information Storage and Organization in the Brain." *Psychological Review* 65 (6): 386–408.

Rougier, Nicolas. 2007. Biological Neuron Schema. https://commons.wikimedia.org/wiki/File:Neuron-figure-fr.svg.

Roy, Ambrish, Alper Kucukural, and Yang Zhang. 2010. "I-TASSER: A Unified Platform for Automated Protein Structure and Function Prediction." *Nature Protocols* 5 (4): 725–738.

Rumelhart, David E., Geoffrey E. Hinton, and Ronald J. Williams. 1986. Learning Representations by Back-Propagating Errors. *Nature* 5 (6088): 533–536.

Rumelhart, David E., James L. McClelland, and the PDP Research Group. 1986. *Parallel Distributed Processing*, volumes 1 and 2. MIT Press.

Russell, Stuart, and Peter Norvig. 2009. *Artificial Intelligence: A Modern Approach*. Prentice-Hall.

Samuel, Arthur L. 1959. "Some Studies in Machine Learning Using the Game of Checkers." *IBM Journal of Research and Development* 3 (3): 210–229.

Sandberg, Anders, and Nick Bostrom. 2008. Whole Brain Emulation: A Roadmap. Technical Report 2008-3, Future of Humanity Institute.

Sanger, Frederick, Steven Nicklen, and Alan R. Coulson. 1977. "DNA Sequencing with Chain-Terminating Inhibitors." *Proceedings of the National Academy of Sciences* 74 (12): 5463–5467.

Scharf, Caleb. 2014. *The Copernicus Complex: Our Cosmic Significance in a Universe of Planets and Probabilities*. Scientific American / Farrar, Straus and Giroux.

Schleiden, Matthias Jacob. 1838. "Beiträge zur Phytogenesis." *Archiv fur Anatomie, Physiologie und Wissenschaftliche Medicin*, 137–176.

Schreyer, Paul. 2000. The Contribution of Information and Communication Technology to Output Growth. Science, Technology and Industry Working Paper 2000/02, OECD.

Schrödinger, Erwin. 1935. "Die gegenwärtige Situation in der Quantenmechanik." *Naturwissenschaften* 23 (49): 823–828.

Schultz, Thomas. 2006. "DTI-Sagittal-Fibers." https://commons.wikimedia.org/wiki/File:DTI-sagittal-fibers.jpg.

Schuster, Stephan C. 2007. "Next-Generation Sequencing Transforms Today's Biology." *Nature Methods* 5 (1): 16–18.

Schwann, Théodore. 1839. *Mikroskopische Untersuchungen über die Uebereinstimmung in der Struktur und dem Wachsthum der Thiers und Pflanzen*. Reimer.

Searle, John. 1980. Minds, Brains, and Programs. *Behavioral and Brain Sciences* 3 (03): 417–424.

Searle, John. 2011. "Watson Doesn't Know It Won on 'Jeopardy!'" *Wall Street Journa*l, February 23.

Seung, Sebastian. 2012. *Connectome: How the Brain's Wiring Makes Us Who We Are*. Houghton Mifflin Harcourt.

Sheffer, Henry Maurice. 1913. "A Set of Five Independent Postulates for Boolean Algebras, with Application to Logical Constants." *Transactions of the American Mathematical Society* 14 (4): 481–488.

Shor, Peter W. 1997. "Polynomial-Time Algorithms for Prime Factorization and Discrete Logarithms on a Quantum Computer." *SIAM Journal on Computing* 26 (5): 1484–1509.

Silver, David, Aja Huang, Chris J. Maddison, Arthur Guez, Laurent Sifre, George van den Driessche, Julian Schrittwieser, et al. 2016. "Mastering the Game of Go with Deep Neural Networks and Tree Search." *Nature* 529 (7587): 484–489.

Sipser, Michael. 1997. *Introduction to the Theory of Computation*. PWS.

Smullyan, Raymond. 1980. *This Book Needs No Title: A Budget of Living Paradoxes*. Simon and Schuster.

Söding, Johannes. 2005. Protein Homology Detection by HMM–HMM Comparison. *Bioinformatics* 21 (7): 951–960.

Solomonoff, R. J. 1964. "A Formal Theory of Inductive Inference," parts I and II. *Information and Control* 7 (1): 1–22, 7 (2): 224–254.

Sporns, Olaf. 2011. *Networks of the Brain*. MIT Press.

Stephenson, Neal. 1995. *The Diamond Age: Or, A Young Lady's Illustrated Primer*. Bantam Books.

Stiles, Joan, and Terry L. Jernigan. 2010. "The Basics of Brain Development." *Neuropsychology Review* 20 (4): 327–348.

Stross, Charles. 2006. *Glasshouse*. Penguin.

Stuart, Greg, Nelson Spruston, Bert Sakmann, and Michael Häusser. 1997. "Action Potential Initiation and Backpropagation in Neurons of the Mammalian CNS." *Trends in Neurosciences* 20 (3): 125–131.

Sutton, Richard S. 1988. "Learning to Predict by the Methods of Temporal Differences." *Machine Learning* 3 (1): 9–44.

Sutton, Richard S., and Andrew G. Barto. 1998. *Reinforcement Learning: An Introduction*. MIT Press.

Turing, Alan. 1937. On Computable Numbers, with an Application to the Entscheidungsproblem. *Proceedings of the London Mathematical Society* 2 (42): 230–265.

Turing, Alan. 1950. Computing Machinery and Intelligence. *Mind* 59 (236): 433–460.

Ulam, Stanislaw. 1958. "Tribute to John von Neumann." *Bulletin of the American Mathematical Society* 64 (3): 1–49.

Van Inwagen, Peter. 1983. *An Essay on Free Will*. Oxford University Press.

Vapnik, Vladimir Naumovich. 1998. *Statistical Learning Theory*. Wiley.

Venter, J. Craig, Mark D. Adams, Eugene W. Myers, Peter W. Li, Richard J. Mural, Granger G. Sutton, Hamilton O. Smith, et al. 2001. "The Sequence of the Human Genome." *Science* 291 (5507): 1304–1351.

Villringer, Arno, J. Planck, C. Hock, L. Schleinkofer, and U. Dirnagl. 1993. "Near Infrared Spectroscopy (NIRS): A New Tool to Study Hemodynamic Changes during Activation of Brain Function in Human Adults." *Neuroscience Letters* 154 (1): 101–104.

Vinge, Vernor. 1993. "The Coming Technological Singularity: How to Survive in the Post-Human Era." In *Vision-21: Interdisciplinary Science and Engineering in the Era of Cyberspace*. NASA.

von Neumann, John. 1945. *First Draft of a Report on the EDVAC*. Moore School Library, University of Pennsylvania.

Ward, Peter, and Donald Brownlee. 2000. *Rare Earth: Why Complex Life Is Uncommon in the Universe*. Copernicus Books.

Watson, James D., and Francis H. C. Crick. 1953. "Molecular Structure of Nucleic Acids." *Nature* 171 (4356): 737–738.

Werbos, Paul J. 1988. "Generalization of Backpropagation with Application to a Recurrent Gas Market Model." *Neural Networks* 1 (4): 339–356.

Wetterstrand, K. A. 2015. DNA Sequencing Costs: Data from the NHGRI Genome Sequencing Program (GSP). National Human Genome Research Institute. http://www.genome.gov/sequencingcosts/.

Wheeler, Richard. 2011. The Structure of DNA Showing with Detail Showing the Structure of the Four Bases, Adenine, Cytosine, Guanine and Thymine, and the Location of the Major and Minor Groove. https://commons.wikimedia.org/wiki/File:DNA_Structure%2BKey%2BLabelled.pn_NoBB.png.

White, J., E. Southgate, J. Thomson, and S. Brenner. 1986. The Structure of the Nervous System of the Nematode *Caenorhabditis elegans*. *Philosophical Transactions of the Royal Society of London* 314 (1165): 1–340.

Witten, Edward. 1995. "String Theory Dynamics in Various Dimensions." *Nuclear Physics B* 443 (1): 85–126.

Woese, Carl. 1990. Phylogenetic Tree of Life. Wikimedia Commons. https://commons.wikimedia.org/wiki/File:PhylogeneticTree.png.

Wolpert, David H. 1996. "The Lack of A Priori Distinctions between Learning Algorithms." *Neural Computation* 8 (7): 1341–1390.

Wrangham, Richard. 2009. *Catching Fire: How Cooking Made Us Human*. Basic Books.

Zemelman, Boris V., Georgia A. Lee, Minna Ng, and Gero Miesenböck. 2002. "Selective Photostimulation of Genetically ChARGed Neurons." *Neuron* 33 (1): 15–22.

Index